JOHN MACKAY

WILBUR S. SHEPPERSON SERIES IN NEVADA HISTORY

University of Nevada Press   *Reno & Las Vegas*

MICHAEL J. MAKLEY

JOHN MACKAY

Silver King in the Gilded Age

For Randi Lee

Wilbur S. Shepperson Series in Nevada History
Series Editor: Michael Green

University of Nevada Press, Reno, Nevada 89557 USA
Copyright © 2009 by University of Nevada Press
All rights reserved
Manufactured in the United States of America
Design by Louise OFarrell

Library of Congress Cataloging-in-Publication Data

Makley, Michael J.
John Mackay : silver king in the gilded age /
Michael J. Makley.
      p. cm.— (Wilbur S. Shepperson series in Nevada
history)
Includes bibliographical references and index.
ISBN 978-0-87417-770-1 (hbk. : alk. paper)
1. Mackay, John William, 1831–1902. 2. Silver mines
and mining—Nevada—History. 3. Capitalists and
financiers—Nevada—Biography. I. Title.
HD9537.U62M335 2009
979.3'02092—dc22   2008041623

The paper used in this book meets the requirements
of American National Standard for Information
Sciences—Permanence of Paper for Printed Library
Materials, ANSI/NISO z39.48-1992 (R2002). Binding
materials were selected for strength and durability.

University of Nevada Press Paperback Edition, 2015
ISBN 978-0-87417-994-1 (pbk. : alk. paper)

Contents

List of Illustrations   vii

Preface   ix

Prologue   1

1. *Starting Out*   8

2. *The Fountainhead*   16

3. *Hale & Norcross*   35

4. *Competitors*   48

5. *The Con. Virginia*   58

6. *Bonanza Silver and Kings*   73

7. *1875*   83

8. *Responses*   101

9. *Status and Scandal*   119

10. *To Try Fortune No More*   134

11. *A Changing Cast*   146

12. *The New War*   159

13. *Near Disaster*   174

14. *Losing Control*   191

15. *Universal Eulogy*   204

Notes   213

Bibliography   247

Index   257

List of Illustrations

(following page 158)

John William Mackay

Louise Hungerford Mackay

The Mackay house in Virginia City

James G. Fair

James C. Flood and William
Shonessy O'Brien

The French Mill

A view of Virginia City

c Street

The Con. Virginia's shaft house

Miners

Comstock's "fissure vein"

Leaving the Bonanza mines

Louise Mackay

Mackay at age fifty

Principals in James Fisk's death

Jay Gould

Cartoon of Uncle Sam
confronting Gould

Mackay-Bennett cable hauled ashore
at Manhattan Beach, New York

Clarence Mackay

John Mackay near the end of his life

Statue of Mackay

Preface

John Mackay regarded Virginia City, Nevada, as his home although he lived two-thirds of his life elsewhere. The sentiment is understandable, since the silver he took from the town's Comstock Lode made him one of the richest men in the world. Moreover, because he rose from the ranks of the miners he was always considered "one of the boys," and owing to his honesty and generosity, when he succeeded he was lionized in his adopted state.

San Francisco interests ran nineteenth-century Nevada mining operations. Because hard-rock, deep-fissure mining required great amounts of capital, the San Francisco stock market, with its sale of shares and assessments, allowed the mines to be developed. But the story of Mackay on the Comstock is of a local miner overcoming the dominance of corporate San Francisco. Although two of Mackay's partners were San Francisco brokers, it was Mackay and fellow "Comstocker" James Fair who out-maneuvered the San Francisco Bank Ring, and it was miners under their direction who brought the Bonanza silver to the surface.

The importance of Mackay to western history would be significant if his story ended with his mining successes. Other histories and biographies have presented considerable information regarding that episode of his life. But only cursory attention has been paid to his second career: building a world-wide communication network, which featured waging war on national and international communications systems that had undergone horizontal consolidation. In his endeavor Mackay opposed Jay Gould, the most vilified of the group of finance capitalists identified by muckrakers as "robber barons." This book does not attempt a wide-ranging analysis of the ethics of Gould's dealings in big business, a task already undertaken by several historians and revisionists. What is more important to this narrative is Mackay's perception of Gould's management of Western Union Telegraph and the transatlantic cable cartel that Gould controlled. This work is not intended to be a critical study of western mining or a history of the Gilded Age. It does not attempt to analyze the dynamics of the monopolization of the Comstock or the national rise of the large corporation. It is a biography, placing the Mackay story in the historical perspective of the era and illuming his successes and failures as he fought against certain of its dominant entities.

Mackay left few personal papers. It is largely because historian Grant H. Smith collected a wealth of material on him in the 1930s and early 1940s that a more fully developed picture of Mackay can be presented. Smith's cache is stored in boxes at an offsite facility of the Bancroft Library. There are many individuals who need to be acknowledged for assisting with this project, beginning with the staff of the Bancroft Library—in particular David Kessler, who insured that the correct boxes of Smith's materials were available on my visits to Berkeley.

I owe special thanks, as well, to the librarians at the Nevada State Library; the University of Nevada, Reno, Special Collections Department; the Nevada Historical Society; the California Section of the California State Library; the Denver Public Library Western History Department; and the Huntington Library. McAvoy Layne provided insightful commentary on portions of the manuscript; Scott Walker suggested works on economic history; and Frank Kovac, as with other undertakings, provided valuable technical support. Knowledgeable help in gathering photographs was provided by Jaquelyn Sundstrom at the University of Nevada, Reno, Special Collections Department, and Lee Brumbaugh at the Nevada Historical Society. Eric Moody, who has consistently assisted me, and Michael Maher, at the Nevada Historical Society, were especially helpful, as was Juan Gomez of Reader Services at The Huntington Library. Peter Blodgett, Curator of Western Historical Manuscripts at The Huntington, and Jennifer Watts, Curator of Photographs there, took time on short notice to give personal attention to the project. University of Nevada Emeritus Professor Jerome Edwards read the manuscript and provided insights that clarified many portions of it. Ronald James also read it, contributing expertise in western history and an understanding of Scottish, Scots-Irish, and Irish heritage. Professor James contributed many hours, and his efforts are reflected in this final version. At the University of Nevada Press, Victoria Davies, Sara Vélez Mallea, and Charlotte Dihoff were particularly supportive. Once again Sarah Nestor has edited what I thought was a finished product, adding, deleting, and suggesting alternatives to make the book more readable. The text, notes, and bibliography are all substantially improved because of her precise and exhaustive corrections. From the book's inception Matthew Makley asked questions and offered advice, and where it succeeds, it is in large part because of his assistance.

Finally, to say that in writing this work I stood on Grant H. Smith's shoulders is a great understatement, for without his work this book could not have been written.

JOHN MACKAY

Prologue

IT IS SAID THAT DURING his bonanza days in Virginia City, John Mackay visited the home of a man who had disappeared after embezzling a large amount of money. Mackay reminded the heartbroken wife that her husband was absentminded. He urged her to search the man's clothes, as he might have left the money in a pocket. She rummaged through the clothing, in the bedroom, but found nothing. As he left the house Mackay gestured to a coat in the hallway, telling her to look there, too. In one of the coat pockets the woman found nearly three thousand dollars, the amount of the shortage. She looked out the door to tell Mackay the good news, but he was far down the block, keeping to the shadows as if having just committed a burglary.[1]

Sam Davis, Nevada newsman and historian, related the anecdote in a eulogy for Mackay in the *San Francisco Examiner*. The story illustrates traits universally attributed to Mackay: generosity and an extreme reticence at being recognized for it. But there are questions about the tale itself. Did Davis, who was not averse to embellishing a story, recount history or lore? Was the story true, or was it a characterization highlighting Mackay's benevolence? Did Davis perhaps concoct the tale to perpetuate the Mackay legend?

From the early 1870s until his death in 1902, John Mackay was among the richest men in the world. Typically, he has been lauded for his role in developing hard-rock mining and his rise from penniless immigrant to a man of rarely matched material achievement. He also has been portrayed as the outstanding character of the Comstock era. Unlike many of his wealthy contemporaries, Mackay did not suffer public antipathy. Amid rampant commercial corruption, he was seen as an everyman who made good. He worked his way up earning "clean money," so called because it was believed that extracting ore from the earth benefited all while harming no one (environmental destruction went largely unnoticed).[2]

In 1890 Fredrick J. Turner presented his thesis that successive frontiers, terminating in the West, gave birth to a unique American way of thinking and a new, noble culture. Over the past twenty-five years historians, in reexamining the West (and then using similar tools in other regions), have brought to light the moral ambiguity, vices as well as virtues, of the culture that evolved in regions where newcomers intruded into indigenous environments. The emergent perspective emphasizes the convergence of various traditions, including those of Native peoples, Latin Americans, Europeans, Africans, and Asians, and the influence of women as well as men. It also studies the prevalence of racism, sexism, environmental degradation, and cutthroat economics and offers new interpretations of their consequences. The new history challenges the ethnocentric concept that English-speaking white males who dominated in finance or warfare "won the West."[3]

Implicit in this refocusing is the question of the degree to which those previously credited with shaping the West deserve honor. This book argues not only that Mackay is worthy of the status awarded him in Nevada and mining history, but also that he warrants standing as an important business leader fighting the consolidation of power in the Gilded Age. His recognition should include his confrontation of two seemingly invincible monopolies: the notorious "Bank Ring" on Virginia City's Comstock Lode and, in Mackay's second career, Jay Gould, Western Union, and the transatlantic cable cartel. These enterprises were of paramount importance and contributed significantly to the evolving mosaic of American life.

In the first case, Mackay and his associates freed the Comstock from a vertical monopoly whose hold was so complete that Davis observed, "People despaired of ever escaping from its relentless grasp."[4] Defeating it moderated rates in milling, timber, shipping, transportation, and water, precipitating the discovery and development of the ore field that came to be known as the "Big Bonanza." In the second case, at a time when the U.S. Congress was contemplating the regulation or even federalization of the telegraphic communication field, Mackay's battles reduced telegraph rates from 38 to 31.3¢ a word and transoceanic cable rates from an exorbitant 75¢ a word to 25¢.[5] His actions facilitated national and international commerce and eliminated the likelihood of federal intervention. Moreover Mackay, who prided himself on being a common man, was proud that his fight against the cartels promoted public use.[6]

Because he shunned publicity and left few personal papers, a study of Mackay presents difficulties. As with the Davis anecdote, the most readily available primary accounts are not entirely trustworthy. The Far West journalists who wrote about him were the equals of any in the period, but those individuals, including Davis, Dan DeQuille, Rollin Daggett, C. C. Goodwin, and John Russell Young, also helped create the mythic old western history with their tales and hagiographic accounts. Nevertheless, a distillation of the collective accounts reveals consistent, effusive praise for Mackay, the first evidence that he was a singular figure.

In the 1930s and early 1940s, Comstock historian Grant Smith collected an invaluable body of information regarding Mackay. As a young job seeker in the 1880s, Smith had met Mackay long enough to be turned down for a position working in the mines. Smith intended to write a book about Mackay, and the materials, notes, sections of manuscript, newspaper clippings, and interviews with those who had known Mackay remain in several large cartons at the Bancroft Library in Berkeley, California. Instead of the biography, Smith used the mining material he had collected to create his comprehensive study, *The History of the Comstock Lode.* He dedicated the book to Mackay, who had died forty years earlier.[7]

Previous books, the most popular of which is Oscar Lewis's classic, *Silver Kings: The Lives and Times of Mackay, Fair, Flood and O'Brien, Lords of the Nevada Comstock Lode,* feted Mackay and his associates. But Lewis's and the other works have treated Mackay and his partners as equals. Although James Flood, who handled finance in San Francisco, and James Fair, who supervised the mines underground, were invaluable, it was Mackay who held the group together. He owned the majority of shares, controlling ⅜ of the stocks, while Flood and William O'Brien split ⅜ and Fair held ⅔. In an age of deceit and machinations, Mackay earned the sobriquet "the honest miner." And it was Mackay who created the transoceanic cable system.

During his childhood, Mackay had earned a pittance as a fatherless Irish immigrant on the streets of New York. After his father's death, he forsook school to help support his mother and younger sister as one of the gang of boys who hawked penny newspapers. As an adult, he spent nights educating himself. Davis told of calling on him one evening on the Comstock to find him with a schoolbook. Mackay did not attempt to hide it. He told Davis frankly, "I never received much education and I have to put in my leisure hours catching up."[8] Mackay spoke with a stutter and a brogue at a time when store windows advised job seekers that Irish need not apply. Once he garnered wealth and fame, his speech attracted little attention from social commentators. San Franciscan Amelia Ransome Neville said his appearance gave him prominence: "I have seen him step out of the Lick House into Montgomery Street . . . a broad felt hat above his keen eyes and sweeping moustache—to be followed by many admiring glances."[9] Yet he avoided the spotlight, wore no jewelry, and, when in Nevada, typically dressed in miner's garb.

Mackay came to Sierra County, California, in 1851, during the gold boom. But the rich claims had already been staked out, and although he was diligent and learned the intricacies of placer mining, as the years passed he watched profits in the county steadily diminish. By the late '50s, with the area having been reworked several times, Mackay ended up felling trees and framing timbers for someone else's drift mine. A fellow who worked with him commented, "Mackay worked like the devil and made me work the same way." Mackay's work ethic never changed. Even when his ownership of the Con. Virginia (Consolidated Virginia) bonanza made him the richest miner in the world, earning $300,000 a month, Mackay rose each day at 4 AM to inspect tunnels.[10]

Although he later lived in San Francisco and New York, Mackay always maintained that he was a resident of Virginia City, Nevada, and he had a special affection for its residents. He provided pensions for many old Comstockers. When Dan DeQuille could no longer support himself, word carried to Mackay, and the millionaire directed that DeQuille be paid an allowance equal to his former salary for the remainder of his life. "What's money for," he commented, "if you can't use it for a friend?" When asked about generous gifts to those he did not know, he said, "They are suffering, that is sufficient."[11]

His gift giving was extensive; however, it was done without a plan. He provided stipends to hospitals, orphanages, and old acquaintances yet failed to create an organization to continue the philanthropy after his death. This led the *New York Herald* to eulogize that in every state there would be some widow, once-forgotten orphan, poor soldier, broken-down miner, failed contractor, needy student, or priest who would miss his charity.[12]

Mackay's business acumen matched his generosity. The fields of economic and business history emphasize the role of improved technology and managers' efficient choices that gave certain early-day firms a competitive edge, allowing them to prevail during the rise of corporations. Economic sociology has argued that power was a more correct foundation in corporate development and competition.[13] Mackay provides evidence for both theories. While consistently hiring competent and experienced managers and employing cutting-edge technology, he utilized resources efficiently in building his businesses. At the same time, once established, he wielded his vast financial resources to engage opponents at a time when social Darwinism preached the survival of the fittest.

While his business career was successful and fulfilling, his marriage proved less so. Mackay met his wife in Virginia City and remained married for the rest of his life, but the marriage was a peculiar arrangement. Although he built a house for her on the Comstock and bought one in San Francisco, for most of their marriage Louise Mackay lived in Europe, while Mackay attended his businesses in America. His lifestyle was unassuming; hers involved social climbing and conspicuous consumption.

Rather than filling the prescribed role of domesticity for wives of the era, Mrs. Mackay endeavored to attain status. Her rise in European society was as spectacular as her husband's in mining and commerce. She used her husband's wealth and her charm to move from the mining camps to topmost status in the social world. Her homes in Paris and London were palatial estates where she entertained lavishly. Mackay was disdainful of the nobility that frequented the houses, and on visits voiced his opinion.

But he loved being with his children. Alexander O'Grady, a member of the San Francisco Bar, who as a child lived with the Mackays while his mother oversaw the Paris house staff of thirty, said that when Mackay arrived, "all business cares and worries were thrown off like a cloak." And, according to O'Grady, husband and wife were affectionate. When one of Mackay's frequent letters from America arrived, Louise would carry it to O'Grady's mother: "I have another letter from my Johnny boy; let me read it to you."[14]

Like her husband, Mrs. Mackay gave generously to charities. During one severe Paris winter, she bought so many warm clothes to be given to the poor that they filled the mansion's attic. In spite of her good will, the people back in Nevada were resentful of her neglect of her husband. Smith noted, "They could not forgive her for taking away the children. Mackay's lonely life was to them a constant witness of a woman's ingratitude." Mackay's Comstock supporters ridiculed how Mrs. Mackay carried on like European royalty, but some who knew her from the early days would come to her defense. Leading Nevada citizen "Old Abe" Curry commented that Mrs. Mackay "looked like a queen in a calico dress." A San Francisco socialite who knew her in the early 1870s concurred, saying neither sudden wealth nor California society had overwhelmed her, and that an easy, graceful independence ensured her social success abroad.[15]

Separation from his family was only one of Mackay's tribulations. He survived an assassination attempt by an elderly man who had lost his fortune. The man, believing Mackay and his partners were responsible for his losses, shot Mackay in the back. Once he recovered, friends urged Mackay to hire bodyguards to protect himself. He laughed, refusing to entertain the idea. He also faced charges of adultery in a widely publicized case but was acquitted when the accusers, a husband and wife, were revealed as blackmailers.[16]

The case had caused particular surprise, because Mackay was renowned for his gentlemanly demeanor. He could, though, exhibit a fiery temper. After a particularly contentious meeting, elements of the press accused him of "hotheadedness" and being a "bulldozer." Mackay's exercise regimen included boxing. He brawled with a notorious pugilist who had once killed a man in a fight and, at age sixty, he pummeled an equally aged, decorated ex-soldier. There were also the inevitable court cases over business and the myriad battles engaged in when gaining and holding a fortune.

Mackay suffered his greatest sorrow when he lost his twenty-five-year-old first son in a horseback-riding accident. At the time grief incapacitated him. For five years he refused to attend any place of entertainment, and friends said he never fully recovered.

It had been many years earlier when Mackay and his associates won their first major battle on the Comstock. They took possession of the contested Hale & Norcross mine from the feared capitalist William Sharon and his infamous Bank Ring. Sharon, in his Virginia City Bank of California office, summoned the upstart. They greeted each other across the rail: Mackay, erect, compact, wary; Sharon, small and frail but with cold, gray eyes reflecting fearlessness. Without retiring to his private office, Sharon broached the subject. He suggested that since Mackay had no experience with such things, Hale & Norcross ore still should be refined at a Bank Ring mill. Mackay replied that different arrangements were being made. The decision would double Sharon's loss. For one of the few times in public, the calculating banker lost his temper. He pointed toward the pass leading out of Virginia City and shouted, "The time is coming when I'll make you pack your blankets back over Geiger Grade." Mackay, always slow to speak, considered the statement. "W-well, I can do it," he said quietly. "I packed them in."[17]

CHAPTER 1

Starting Out

JOHN MACKAY was born in Dublin, Ireland, on November 28, 1831. Many years later, when he was among the richest men in the world, Alexander O'Grady quoted him as saying that he came from a family of "bogtrotters" and used to go barefooted as a boy, "with a pig in the house." Mackay may have said this for effect rather than as autobiography, since his audience included European gentry, of whom he was often contemptuous. His wife, Louise Mackay, commented: "He hates aristocratic pride and prejudice, and has not a high idea of aristocratic virtue." All ties with relations in Dublin were severed when the family immigrated, and they were never reestablished. After making his fortune, he visited, but on his return he commented only that the beautiful green fields were unchanged and that there were no friends left there. His wife said: "He has no relations that he knows of in Dublin; at any rate, none who fixed themselves by kind attentions in his memory when he was a boy."[1]

Mackay's Scots-Irish father and Irish mother brought him and his younger sister to America in 1840.[2] The family lived in New York City on Franklin Street, not far from City Hall Square. Mackay's first days in America were filled playing sports with other children at City Hall Park, which later became the New York City Post Office grounds. Although not pugnacious, he occasionally fought because of a temper that a *New York Times* obituary graciously labeled as the "assertion of a sturdy personality."

It was reported that he knew the city's millionaires by sight, and he told of an early vision that remained with him all his life: his hero, James Gordon Bennett Sr., "hurrying through City Hall Square in New York, as I played there as a boy, a man with a hurried step and a bundle of newspapers under his arm." Bennett was the founder of the *New York Herald,* and many years later his son would become Mackay's partner in the great Atlantic cable venture.[3]

Mackay's father died within two years of their arrival in North America. Mackay seldom attended school afterward, needing to bring in what money he could, selling newspapers and doing odd jobs. His sister, when old enough, entered a convent, where she remained all her life, eventually rising to the position of mother superior. In his teens Mackay was apprenticed to shipbuilder William H. Webb.[4] Another consequence of his father's premature death was Mackay's diminished identity with the Scots-Irish half of his ancestry. He grew up an Irish Catholic, although friends never knew him to attend church as an adult.[5]

The country was short of labor in the 1830s, when canal fever struck and the first wave of Irish Catholic laborers arrived. (Scots-Irish Protestants, as well as a few Irish Catholics, had been in America since at least 1713.) The canal building featured digging through swamps such as the Montezuma marshes in Syracuse, New York, where "the Irish toiled away naked except for shirt and cap among formidable clouds of mosquitoes." The canal diggers received a bad name, when hatred between Scots-Irish and the Irish Catholics boiled over in Indiana on the anniversary of the Battle of the Boyne and militia had to be called in to prevent a pitched battle. In 1834 federal soldiers were summoned to intercede in a similar clash between Scots-Irish and Irish Catholics working on the Chesapeake and Ohio Canal.[6]

In the 1820s an average of less than 10,000 immigrants entered the United States annually. Between 1830 and 1840, the year the Mackays arrived, with just over 17,000,000 in total population, 500,000 newcomers entered.

Approximately 175,700 were Irish. In the 1840s over 1,500,000 people immigrated, of which 666,145 were Irish. And Irish immigration reached its peak in the eight years after the great famine of 1846, when approximately 150,000 people a year made the journey. Some newcomers earned the natives' scorn by immediately applying for public assistance, but American working families' hatred of those who came to compete for jobs ran even deeper. They refused to work with the new arrivals, who were thus limited to manual labor. With racism endemic, many newcomers were shut out completely, and street fights between established citizens and immigrants were common. Catholic churches and even convents were attacked and, at times, destroyed. In the early 1850s the hatred led to the formation of the viciously anti-Catholic Know-Nothing Party.[7]

Prejudice against the Irish persisted throughout Mackay's lifetime. Fifty years later, when he was among the wealthiest men in the world, a Scot said of him: "Mr. MacKay is nae mair an Irishman than I am. Because a cat's [born] in an oven is it a baker's pie?"[8] But in his youth Mackay was confronted with the intolerance directed at Irish Catholic immigrants. In the midst of New York City, he learned to work hard and, perhaps owing to his stutter as well, keep his mouth shut. In the Webb shipyard he kept his temper under control, ignored the politics of race, and learned to build with wood.

The Webb family was one of a handful of renowned companies in the world building clipper ships. The industry would construct 160 of the swift ships within four years, the discovery of gold in California the impetus. Beginning in 1849 in the United States, practically all that was talked about was gold and those leaving to seek it. Annually for four years, 50,000 men left for the unknown land that promised heretofore-undreamed-of rewards for industry.[9] In 1851 Mackay completed his shipbuilding apprenticeship and joined the rush to California.

It has been suggested that Mackay sailed on one of the ships that he helped build. Whether or not this was true, it is probable that he worked as a crewman for his passage. A pressing failing of the clippers, along with taking on considerable water and being uncomfortable for passengers, was their need to maintain large crews. There are differing accounts of Mackay's route: sailing around Cape Horn or traveling by way of the Isthmus of Panama.[10] The choice was surviving the hundred-odd day voyage around the Cape on boiled beef and flour duff or risking cholera and yellow fever while traversing the isthmus by train to wherever the slowly advancing railhead had reached and then straggling through the jungle by mule train to catch a northbound ship for California. In either case, Mackay was one of many thousands arriving in San Francisco by ship that year. The population of California, 117,500 in 1850, was 264,400 by the end of 1852. With board in a hotel or tent from twenty-five to forty dollars a week, Mackay headed immediately for the goldfields, first to the American River, then to Sierra County.[11]

When Mexico controlled California, the term "sierra" referred to two or more mountain peaks in a row. In 1852 in northeastern California, tiny, mountainous Sierra County was created, extending fifty miles east and west by twenty miles north and south. On its eastern boundary, Bald Peak stands at 8,749 feet, falling to the west to the 5,000-foot Sierra Valley. From the valley the land drops to below 3,000 feet at Downieville, at the fork of the Yuba River, some ten miles from the county's western boundary. In spring 1852 the area was teeming with people, many brought there in 1849 and 1850 by the fabulous story of a lost gold lake, where coarse gold could be scooped up by the handful. The early immigrants were generally ignorant regarding gold deposits, and many strange geological theories evolved, including the most popular theory that somewhere in the high mountains was the lake that was the source of all the gold.[12] Mackay, twenty years old, arrived in the hub of mining activity after the fabled lake theory had been debunked by several extensive, unprofitable expeditions. He set up camp at Durgan's Flat, near Downieville, and worked the gold-bearing forks of the Yuba and its innumerable smaller streams, bars, and subterranean beds.

As might be guessed, most miners shared the trait of restlessness. Frequent reports of new strikes, where riches were easier to come by, initiated great migrations. In 1851 stories were spread that the sand at Gold Bluff in Klamath County was laden with gold. In 1853 it was rich gold mines found at the headwaters of the Amazon in Peru. In 1855 the Kern River strike in central California caused more excitement than any that preceded it. In 1858 there was news of deposits on the banks of the Fraser River. With each account and rumor hundreds of miners gave up the claims they were working to pursue the new dream.[13]

Mackay was unaffected by the feverish speculation. There was ore being found around Downieville, as well. Miners occasionally struck the famous "blue lead," an ancient channel beneath sheets of lava whose fragments were choked with auriferous gravel. Thirteen miles east, in the buttes, a ninety-seven-pound troy weight nugget was discovered, and at the Eureka digs to the west, a one-year annual cleanup netted $1,400,000 at the six-hundred-foot level.[14] Mackay prospected the area for seven years, learning about placer and quartz mining and the geology of the Sierra Nevada. Fellow miner Rollin Daggett, later a Nevada newspaper editor and U.S. congressman, commented: "John Mackay [undertook] a great deal of hard work as a miner on the North Fork of the Yuba River at Downieville, where many a day he stood in the muddy stream up to his hips, tending a mining flume."[15]

After the swarm of 49ers engaged in their wild rush to Sierra County, mining rules were formulated and social organization allowed the formation of communities. Camps flourished under camp law, some continuing under their own rules long after California counties were formed and county governments established.[16]

There was a kinship among individuals in the first waves of immigration. Miners, 125,000 strong by 1851, acknowledged no social classes. The exception was the universal racist views regarding Native Americans, African Americans, Mexicans, and Chinese, who were almost always excluded. Gold mining was intended to provide opportunity for the Anglos from the East, not minorities. In 1850 the California legislature levied a tax on any miners of non-English descent, in particular the Mexicans, French, and Chileans. Among whites lineage could not be proved, so hierarchy was determined only by success (with nearly $76,000,000 produced in 1851, there was plenty of it), and everyone, regardless of age or education, began as equals. The prejudice against Irish Catholics, so prevalent in eastern cities, was largely unrecognized in the gold-mining towns. Preserving order and getting along were voluntary duties taken up by almost every citizen.[17]

Sierra County had several thousand residents in the 1850s. The roads into it were not wide enough for wagons, so mule trains, some as long as seventy-five animals, carried supplies. A sense of the county's commerce is suggested by the traffic between the regional hub, Marysville, and Downieville; operations required twenty-five hundred mules and employed between three and four hundred men. One merchant described the travails of carrying goods into Sierra County. His train usually consisted of thirty-five mules—six for riding, the rest carrying 300 to 425 pounds of flour, sugar, beans, whiskey, bales of clothing, and mail. A typical trip, making sixteen to eighteen miles a day, took four days. Carrying on transportation in the winter was more difficult, with trains often starting at one or two o'clock in the morning, when the snow had become crusted and hard, and halting when the sun rose high enough to again soften it. Midwinter snows that reached thirty feet deep on the passes shut down all transport except that carried by twelve-foot-long skis, known as "Norwegian skates."[18]

The county became an extreme example of a mining community, progressing in fits and starts toward a civil society. Saloons, gaming tables, and houses of prostitution were run twenty-four hours a day, and a tolerant attitude toward them was the general order. The Gem Saloon at Durgan's Flat, where Mackay lived, is an example of how such an institution might gain community acceptance. It was a matter of remark that it contributed heartily to the support of the local preacher, from no other motive than decency.[19] Mackay, sober and steady, was a model citizen in the county. He was well liked but not overly social—continuing to swing a pick when others had retired to their cards and drink. Some of the others' difficulties gave the community a widespread reputation for enacting mining-camp justice.[20]

William Downie, for whom the town was named, described an early instance of the enactment of miners' law. On the Fourth of July 1850, alcohol fueled carousing and hilarity. In the afternoon two men, their brains "giddy," quarreled and fought. One, a newcomer, pulled a knife and stabbed the other. The wound, although spurting blood, would not prove life threatening, but at the time the blood fueled outrage. The miscreant was immediately brought to trial, the sentence thirty-nine lashes on the bare back. "Big" Logan, a cousin of the wounded man, wielded the whip and, as each blow drew more blood, "Logan seemed the only man in the crowd who was entirely unmoved by the horrible spectacle." Downie commented that the flogging served its purpose, because thereafter if a knife was pulled the Fourth of July incident was brought up, and "the effect of such a reminder was simply magical."[21]

One year later, another stabbing resulted in action so shocking that it was denounced in scathing terms in the *London Times* as an example of American lawlessness. The offender was a woman. Juanita, who was of Mexican descent, lived with a gambler and worked in a gambling saloon. On July 5, 1851, after a speech by a U.S. senator, miners again celebrated—this time far into the night. In the early morning hours, a popular man in the territory broke down the door of Juanita's residence. He later returned to the house, and she stabbed him to death. That day what one observer called "the hungriest, craziest, wildest mob that I ever saw" placed her on "trial." The man's friends contended that he had only returned to apologize, and despite the claims of Dr. C. P. Aiken that she was pregnant (two other doctors examined her and concluded she was not), Juanita was convicted. Saying she would repeat the act under similar circumstances, Juanita secured the noose and stepped off the bridge.[22]

Between 1852 and 1856 half a dozen newspapers were started in Downieville, including one short-lived temperance sheet. Fraternal societies were prevalent. In 1858, at Forrest City, Mackay joined the most prominent of the societies, the Free and Accepted Masons, established as a California lodge in 1850. Brotherhoods looked after their own and dispensed charity to the needy, but when it was necessary to raise money for general purposes, the entire community was tapped. On an occasion when money was owed teachers, a play was staged, with young men taking women's roles, bringing in seven hundred dollars in one night. When a new school was needed, it was funded by a miners' subscription list. By the mid-1850s Downieville was surpassed in population in California only by San Francisco, Sacramento, and two other mining towns, and it missed becoming the state capital by one vote. Men with money developed the hard-rock mines, floating bucket operations, and later the hydraulic diggings. There were forty prosperous quartz mills and thirty sawmills in the area.[23]

Mackay earned enough to live on, and to support his mother back in New York. By all accounts he was steadfast in his approach, using forethought, and was a master of detail. Part of his legend was that he found gold where others believed they had exhausted the pocket. But his efforts did not yield consistent success. Sierra County's mining wealth declined steadily over the time Mackay worked there. The average amount of gold taken by placer mining went from twenty dollars a day in 1848 to five dollars in 1853 and three dollars by 1860. When times were hard, Mackay worked for wages until he could strike out on his own again. On at least one occasion, he and his friend Jack O'Brien took day-labor jobs lumberjacking and framing mine tunnels. In 1858 Mackay, O'Brien, and four other miners lived together in a log cabin, and Mackay said, "I used to figure on every cent."

The easygoing O'Brien was Mackay's best friend. Socially he was quite different from Mackay, being convivial and witty. Long after O'Brien's death, Mackay said, "Jack was a great gentleman and I loved him better than any man I have ever known." In later years Mackay looked on those times as the happiest of his life.[24]

There was a drugstore in Downieville called "At the Sign of the Pestle and Mortar," which, besides drugs, sold iced creams and soda. Major Daniel E. Hungerford, who had marched on Mexico City with Winfield Scott, ran it. His daughter Louise was fifteen and a half in 1859 and was engaged to a twenty-four-year-old physician, Edmund G. Bryant. Louise had spent a year studying French and Spanish with the Dominican nuns at Saint Catherine's Academy in Benicia, California. She was reported to be pretty, charming, and cultured, with a winning disposition. Although living in close proximity, there is no evidence that she and Mackay met until seven years later in Virginia City, when she was widowed and they became engaged.[25]

At the beginning of June 1859, to the east of the Sierra in a place named Washoe after the Indians who resided there, Peter O'Riley and Pat McLaughlin were mining at the head of a ravine above Six Mile Canyon. They dug a hole to gather water for use in their rockers and discovered a bed of black, decomposed ore. Mixed with it were spangles of gold. But the dark, heavy sand settled to the bottom of their rockers, covering the quicksilver and preventing the amalgamation of the gold. They did not recognize the black, lumpy material, for it did not glitter. Throwing it out of the way, they collected the gold in great quantity. Before long they were taking out flakes worth as much as a thousand dollars a day. A month later it was discovered that the lumpy substance they were cursing and tossing aside was silver.[26]

By August 1859 stories of the strike, soon to be known as the Comstock Lode, carried to Sierra County. This time word of a new strike intrigued Mackay. Concluding that their prospects in Sierra County had played out, he and O'Brien packed their blankets and started the one-hundred-mile walk to Washoe.

CHAPTER 2

The Fountainhead

THE COMSTOCK LODE was discovered at an elevation of 6,200 feet on the east slope of Mount Davidson, whose summit reaches 7,775. From Mount Davidson's peak, the view to the east is of arid hills and valleys as far as the eye can see. Twenty-five miles to the west the view is blocked by the rugged, snow-capped Sierra Nevada. The massive peaks of the Sierra, rising 10,000 to 11,000 feet, curve from sight in a crescent north and south, steadily gaining height in the south (the Sierra's highest peak, Mount Whitney, rises above Death Valley at 14,494 feet).[1]

Mackay and O'Brien crossed the mountains from Sierra County, arriving at Virginia City in December 1859. Legend tells that when they reached the head of Six Mile Canyon, O'Brien pulled a fifty-cent piece from his pocket, the last of their joint capital, and threw it far down the canyon, asserting that now they might enter the camp like gentlemen.[2]

There were some three hundred men already in the camp, and all the ground on all sides of the original strike was claimed. The lead on the surface, marked by great outcroppings of quartz, stretched north and south, and companies working that ground were finding ore. The original strike was called "Ophir," and the camp was initially called Ophir Diggings.[3] The name was changed to "Virginia," after "Old Virginia" James Finney (or Fenimore), when early miners held a meeting in September 1859 and named the town for him as the original locator of the ledge. After he was thrown from a horse to his death, the people issued a resolution that read in part: "Resolved that while we humbly bow to this dispensation of Providence, we recognize it not less a matter of propriety than duty to give our testimony to the virtues of the deceased; and whilst acknowledging his faults—faults common to mankind—we deem it not complimentary, but just, to say that James Finney was ever known among the people of this territory as a generous, charitable, and honest man."[4]

Finney's faults were attributed in large part to his excessive use of alcohol, and apparently alcohol was the impetus for naming the town. Almarin B. Paul, who built the first ore mill in Nevada, said "Old Virginny" used to visit while he was constructing the mill. Paul quoted Finney as saying: "Well, we laid in, among other things, some whiskey. We all had a liking for whiskey, and going back, lightened up all the bottles except one, and that I held on to. Going along, we got into a dispute about the name of the contemplated city—all of us a little sprung. The dispute was waxing warm, when an unlucky boulder happened to be in my way, and over it I stumbled, and away went the bottle, whiskey and I. That settled the question! There was something ominous about it, and the place was baptized 'Virginia City.'"[5]

When Mackay arrived he obtained a contract from the Union mine, which adjoined the Ophir mine on the north, to drive a tunnel into Cedar Hill. He was to earn an interest in the Union for his work. O'Brien did not wish to join him, so Mackay conveyed Kinney Soit fifteen feet of his interest and Alec Kennedy twenty feet to work with him, their work taking most of the winter. It was at this time that Mackay began to earn notice. Although he would have to wait for his opportunity, his abilities marked him for success early on.[6]

In the middle of November two feet of snow fell. All available wood was used for fuel, so standard housing was a tent or rough rock structure, although blankets, shirts, and potato sacks were also reported as building materials. Many miners made excavations in the mountain's side and lived in what were called "holes in the wall." Mackay, O'Brien, Kennedy, and Pat S.

Corbett, later a U.S. marshal, built a dugout into the hillside above the office of the Ophir. Ironically, the structure was near what would become the California mine, atop what years afterward was the Big Bonanza. Like all the others, the shelter was primitive, partly walled with stone. Poles supported a roof of brush, covered with a layer of dirt and a canvas tarp that secured it. A rough door and a fireplace and chimney completed the structure. Four primitive bunks and stools furnished it.[7]

It was reported that about Christmas five feet of snow fell on the camp, and the men dug further into the mountain. One hole in the wall, on what later became B Street, contained two billiard tables and furnished accommodation for twelve men. As winter progressed the Mackay cave took in two new men, one who did the cooking, the other who earned his keep by providing firewood—earned mostly by making night raids on neighbors' woodpiles, the Ophir being his primary source. The newcomers slept wrapped in blankets on the dirt floor. Their answer to snow blowing through the chinks in the walls was to roll themselves tighter in their blankets.[8]

Toward spring the scarcity of provisions became acute. In the tents that served as restaurants, meals cost two dollars or more. The melting snow caused flooding in the caves, and the miners suffered hunger and deprivation. Years later the *Territorial Enterprise* reported that in the Mackay cabin everyone had spent his money and was lying around famished. Finally, an entrepreneur packed in provisions from Placerville. Mackay had thirty dollars tucked away and used it to buy a sack of flour, bacon, tea, and sugar.[9]

The previous August, about the time Mackay and O'Brien started from Sierra County, 3,151 pounds of black ore were carried from the Comstock to San Francisco. Throughout the winter, silver bars created from it were on display in the window of the Alsop & Company Bank. The exhibit, and fantastic stories of the vastness of the strike, created a contagion of excitement. As soon as the Sierra snow started to melt, the great rush began. Boats from San Francisco reeled under loads of freight as they chugged upriver to Sacramento. Food, clothing, tents, blankets, cooking implements, tools, and alcohol were unloaded at docks to be hauled by wagons retracing the movement of the pioneers along the old emigrant trail. At Placerville hundreds of tons of freight stacked up in the streets as snow blocked the way. Before the road was clear, pack animals started, struggling to pull rudely constructed sleds up the mountain and down by way of Johnson Pass. Behind the cargo, thousands of people gathered. As soon as humanly possible, they struggled through snow, slush, and mud, thronging to the territory.[10]

In a letter to his wife, J. Ross Browne, later U.S. commissioner of mines and mining for Nevada, described the camp in the first week of April 1860, saying there was no other place on earth as bad as Virginia City: "It is perched up amongst desolate rocks and consists of several hundred tents, holes in the ground with men living in them like coyotes, frame shanties and mud hovels. The climate is perfectly frightful, and the water is so bad that hundreds are sick from drinking it."[11]

The conditions were not bad enough to blunt the rush, but in May and June it faltered as word of a war with Pyramid Lake Indians carried across the Sierra. California Native Americans, conditioned through two generations of mission relations, had engaged in few fights with the gold-crazed intruders, but the tribes of the Great Basin had had no mission experience and little contact with whites. Now they were being stripped of their means of survival; emigrants were usurping their game, land, rivers, and lakes. Northern Paiutes, anxiously searching for two missing young girls of the tribe, found them imprisoned and "most brutally treated" by three brothers who owned Williams Station. Tribal members rescued the girls, executed the kidnappers, and burned the station to the ground. Although tribal custom and frontier justice were in accord, William Ormsby, a community leader in nearby Carson City, was attempting to bring what he perceived as order to the territory. Amid wild rumors of what had occurred, he rallied 106 men, many from Virginia City, seeking retribution against the Indians. Underestimating the Northern Paiutes, who were reinforced by Bannock Shoshone, the disorganized attacking party was out-maneuvered. In the battle, near Pyramid Lake thirty miles northeast of the Comstock, Ormsby and seventy-five other whites were killed.

A general alarm spread to Sacramento, San Francisco, and up and down the Sierra. The miners sent women and children back over the mountains or secured them in stone blockhouses and, buoyed by regular cavalry and artillery, formed a militia. Hundreds of men marched to the place of the former battle. A brief skirmish resulted in the vastly outnumbered Paiutes slipping away into the desert. Tribal leaders eventually sued for peace, and the tribe returned to Pyramid Lake. But this incident and a number of smaller attacks by other Indians caused several forts to be established in Nevada and spurred miners in California to begin winter campaigns of annihilation against that state's Native peoples.[12]

At about the time that it was determined the Indians near the Comstock were not seeking to indiscriminately attack whites, the populace realized that the rich ore clustered around the original strikes was not spread universally throughout the area. This further stymied the rush, but over the summer it regained momentum as other strikes were made.[13]

In spring Mackay's Union tunnel reached the lode, where they found nothing but masses of barren quartz. The tunnel a failure, Mackay went to work in the Mexican mine (originally called the "Spanish"), framing and installing timbers. The Mexican, a one-hundred-foot strip two hundred feet from the south end of the original Ophir strike, was being systematically dug and would be one of the Lode's best producers during the next couple of years. Mackay was working for the Maldonado brothers, who, along with nine associates from Mexico, were developing the area. It is worth noting that he was working for and with Mexicans at a time when the approximately one hundred Hispanics living on the Comstock were apparently relegated by Euro-Americans to a designated living area.[14]

As depth was gained in the Mexican and in the Ophir next to it, water began to flow into the shafts. Small seven-foot-high tunnels were cut into the hill below to act as drains. The drain tunnels, although not needing extensive timbering, required posts and caps to hold the roofs in place. Many years later the *San Francisco Call* stated, "While other timberers were receiving but $3 a day for their work [Mackay] was paid $6." Whether true or not (no other source mentions his collecting double pay), it emphasizes the point that, owing to his shipbuilding training and work in Downieville, Mackay was one of the most skillful timber workers on the Comstock. The Mexican advanced to a depth of fifty feet in April 1860. Yellow clay that held nests of silver in the shape of "flattened wires" was being taken out. The thin, malleable silver was coiled and when unrolled was up to a foot in length, often with small branches extending from it. Some of the "wires" had a bright silver luster; others were blackened; all was nearly pure silver.[15]

The first great mining strikes in present-day Mexico were made in the sixteenth century. When gold was first discovered in California, the Euro-American miners consulted with the Mexicans regarding methods and then adapted them. On the Comstock a dozen years later, most mines were conveying rock to the surface using systems of windlasses and buckets, or having miners push handcars. But the Mexican mine's ore was hauled to the surface in the traditional Spanish system, using rawhide sacks carried by straps around the forehead. Over the summer the mine was developed to the 160-foot level.[16]

By early fall a serious problem arose in the Ophir and the Mexican that would reoccur in all Comstock mines. The clay and decomposed porphyry walls that surrounded the ore were extremely unstable. The ore was often soft and crumbly, and the clay expanded when air was introduced by tunneling. The pressure caused the whole area to swell and move. Adolph Sutro, who years later built the Sutro tunnel, noted in a letter to the *San Francisco Alta* on April 20, 1860, that there was no system by which the mines were working; that instead of running a tunnel below their workings and then sinking a shaft to meet it—alleviating ventilation and water-drainage problems, they were digging large openings that required massive timbering. Most mines were using 30-foot-long round logs supported by close-set pillars, with 18-foot log caps running across the tunnel. The Ophir, at the 175-foot level, came upon a ledge of soft ore 65 feet in width. It was too crumbly and much too wide to be supported by the timber system.[17]

Stymied in trying to extract the ore, about the first of November a director of the Ophir called upon Philipp Deidesheimer, a twenty-eight-year-old practical-mining engineer working in El Dorado County, California. The German native was sent to Virginia City to study the Ophir problem. Throughout November Deidesheimer experimented, and in early December he struck on the solution—an invention he called square sets. Each set was a quadrangle of 12-inch-square posts, 6 feet high and 5 feet long. The upper end of each post was notched so that four additional posts might rest upon it and another could rise, notched into its center. In this fashion the sets could be expanded in any direction. Wedges were used to secure each set against the wall or roof it supported, and the ore could be successfully mined from wall to wall.

Deidesheimer did not patent his invention but freely shared it, and soon all the major mines were using it, allowing the entire Comstock Lode to be tapped to depths of thousands of feet. The exception was the Mexican, which was using a Spanish system developed in the sixteenth century of timbering while leaving pillars of ore as additional supports. Mackay had not worked in it for a couple of years when, in July 1863, the Mexican collapsed from the surface to the 225-foot level.[18]

Virginia City was in continuous flux in 1860. By August wagons—from luxurious Concord coaches to outmoded Conestogas, riders on horseback or mules, and pedestrians in unprecedented numbers were crossing the Sierra. On a single day 353 wagons were counted moving laboriously over the road, with another 50 reloading in Sacramento and 50 more preparing to return. Lengthy lines caused delays on steep mountain grades. It took three long days to travel from Sacramento.[19]

J. Ross Browne explained the layout of the town the newcomers found upon arrival:

> [Virginia City] lies on a rugged slope, and is singularly diversified in its upris-
> ings and downfallings. It is difficult to determine, by any system of observa-
> tion or measurement upon what principle it was laid out. My impression is
> that it never was laid out at all, but followed the dips, spurs, and angles of the
> immortal Comstock. Some of the streets run straight enough; others seem to
> dodge about at acute angles in search of an open space, as miners explore the
> subterranean regions in search of a lead. The cross streets must have been
> forgotten in the original plan—if ever there was a plan about this eccentric
> city. Sometimes they happen accidentally at the most unexpected points; and
> sometimes they don't happen at all where you are sure to require them.[20]

Sparse forests of piñon pine trees, having adapted through the centuries to dry, windy conditions by growing innumerable branches but reaching no taller than twenty-five feet, covered the hills around Virginia City and the adjoining town of Gold Hill—just over a two-hundred-foot rise known as "The Divide." The birds and animals dependent on pine nuts were banished as the miners cut the trees for fuel. When industrious Chinese woodsmen dug out the stumps, the hills were left barren except for the hearty sagebrush.

That first summer thousands of people came to the Comstock. Over half viewed the prospects and immediately retraced their steps. Still, in Virginia City as well as in Gold Hill, tent hotels, general stores, and saloons sprang up. The area was a tumult of activity. Wooden buildings, using pine from forests in the neighboring Sierra, were being built to replace the tents. New industries sprang to life. Before winter there were 100 businesses and 868 dwellings in Virginia City, with something less than 4,000 people in the vicinity.[21]

Ore mills were erected, and milling equipment hauled in. Miners and mill hands were sent for, and by the end of the first year mills hummed and pounded, while piles of waste rock began accumulating below the mine shafts. To record all that would occur, newspapers began publishing. The *Territorial Enterprise* moved from Carson City and, between 1860 and 1875, exerted as great an influence as any paper in the West.[22]

Virginia City's streets were stacked atop each other, traversing the steep mountainside. They were decidedly narrow because of the lack of space, and the buildings above overlooked the rooftops of those below. Some years later the visiting Mrs. Frank Leslie, whose husband published the popular *Leslie's Illustrated Weekly,* expressed the disquieting thought that "the adhesive power may become exhausted, and the whole place go sliding down to the depths of the valley below."[23]

The main business section, initially on B Street but soon centered below it on C, featured commercial buildings, restaurants, laundries, liveries, dance halls, gambling parlors, and saloons. Mrs. Leslie commented that profanity was "fearfully prevalent" in these last structures, which remained open through the night. Above B Street were homes on steep and barely passable lanes.

Below C Street, on D Street and on haphazard, twisting streets labeled E and F, were Chinatown, mine entrances, work yards, and small cabins—a number used by prostitutes. South C Street was a violent, crime-infested area known as the Barbary Coast. It had six unsafe saloons and several houses of prostitution, and the line between them often blurred. Out of 2,200 women who were listed in the 1870 census, 160, or 7.3 percent, listed their occupation as prostitution. Walking at night (accompanied by two police officers), Mrs. Leslie commented: "Every other house was a drinking or gambling saloon, and we passed a great many brilliantly lighted windows, where sat audacious looking women who freely chatted with passers-by or entertained guests within." The *Annual Mining Review and Stock Ledger* said that the territory wore its worst side out. Rollin Daggett commented: "It was a rough camp and the boys were generally pretty wild fellows." Throughout the area men carried guns and discharged them at frequent intervals. Comstock lawyer Zinc Barnes said, "Colt's revolver did more than the Declaration of Independence to make men equal."[24]

A newsman of the era criticized quartz mining. He pointed out that placer mines were the mines of the people, as placer miners relied on their own resources to attain success. The hard-rock miners, toiling as drones deep in the earth, worked for day wages, while corporations provided the attendant expenses so to reap the great profits. American placer miners were rarely content to work long for mining corporations. Their motivation was to make their fortune. They might work at hard-rock mining, but only to raise a stake so they could move on to distant placer claims of their own. Mackay had done that the years he mined in Sierra County, commenting to Jack O'Brien that his goal was to earn twenty-five thousand dollars: "That is enough for any man; with that I can make my old mother comfortable." On the Comstock he changed his mind. He began pursuing ways to develop his own deep, hard-rock mine.[25]

In 1861, having accumulated a small amount of capital, Mackay went with John Henning to Aurora, another boomtown in western Nevada. Henning was associated with future Nevada senator William M. Stewart and two others in the profitable Morgan Mill on the Carson River. In Aurora, a bustling camp of several thousand, Henning and Mackay bought the Esmeralda Claim. Although some Aurora mines seemed at the time to rival those of Virginia City, they would play out within a few years. The Esmeralda soon proved a failure for Mackay, and he returned to Virginia City.[26]

Over the course of the next two years, the Comstock developed above ground and below. As a matter of course, Californians with money bought out the original claim holders and began to sink deep shafts. The Ophir led the way. Three thousand tons of selected ore, milled in 1862, yielded $326 a ton despite losing at least one-third in the tailings. Its wealth continued to the three-hundred-foot level, where the ore began to narrow and decrease in value, before ending at the five-hundred-foot level.

The Mexican worked good ore but lost the vein at three hundred feet. Its owner borrowed $170,000 to build an extravagant mill and, unable to repay the money, lost the mine just before it collapsed on itself in 1863. Shortly thereafter the mill burned down.

Early in 1859, across The Divide, the Little Gold Hill group began digging. These were placer miners, each claiming from eight and three-quarters to fifty feet of disintegrated quartz across the surface of a small, flat-topped hill. A few feet below the surface was the Old Red Ledge, a body of ore extending nearly five hundred feet, from ten to twenty-five feet in width. Three of the original locators retained ownership of their portions; the others sold to incoming Californians, receiving an average of fifty dollars a foot. The larger claims were quickly subdivided, and twenty small, independent owners amicably worked the mines, often sharing shafts. Although substantial fortunes were taken from the various mines, most of the group quickly spent what they made, and nearly all died poor.[27]

At the highest point of the lode there were large outcroppings of quartz, which often accompanied gold and silver, but little ore. Two mines adjoining each other were bought and combined to form the Gould & Curry. Early work in the mine went slowly, with no results. The group of investors included George Hearst; Lloyd Tevis, future president of Wells Fargo & Company; John O. Earl, shipping magnate; and several men who were afterward associated with the infamous Bank Ring. At the end of 1861, one thousand feet below the original diggings, a tunnel was begun from D Street. It struck forty feet of solid ore. This strike reinvigorated owners of mines who as yet had had no success.

Robert "Bob" Morrow, a shrewd and aggressive superintendent, got permission to dig south from the Gould & Curry's D Street tunnel. As he suspected, the ore body extended into his mine, the Savage. He did not finish sinking his own shaft until April 1863. For the next two years he worked in pay ore, taking out ore worth over $3,600,000 (although half the sum was spent in milling).

Late in 1861 the Potasi mine discovered an ore body that dipped into the Chollar mine, located next to it. The Chollar filed suit. Later the Chollar found a body that dipped into the Potasi, and the Potasi sued. Litigation between the two went on for four years. Ultimately the courts ordered that the two mines be consolidated, but not until the proceedings caused three judges of the Nevada Supreme Court to resign over charges of corruption.[28]

While healthy bodies of ore were worked in this handful of mines, 23 other mines returned little or nothing—although substantial amounts of money were sunk into them. There were also nearly 500 smaller wildcat mines throughout the district in 1859, a number that rose to 1,390 the next year.[29]

Mackay was busy. He was earning ownership in properties by doing development work and raising money by digging or setting timbers to buy interests in others. Between February 1860 and February 1864, he traded twenty-seven claims. Typical of his actions was his agreement with James Keefe, who would deed 225 feet of the Milton mine, west of the Chollar, if Mackay ran a tunnel until striking ore. On November 25, 1862, the feet were deeded to Mackay, the conditions of the contract having been fully complied with. This was done, for the most part, while Mackay was working for wages in another mine.[30]

The art of industrial hard-rock mining was perfected on the Comstock and then carried to mines around the world. In the early days everything was trial and error. Below ground Mackay, like all the others, toiled by the dim light of candles, digging through rock. He labored at hand drilling, swinging a hammer or pick, and mucking the loose rock out. And he timbered the tunnel, gauging the strength of the walls to decide the extent of timbering needed. He breathed bad air, suffered in heat that rose one degree for every thirty-three feet in depth, risked floods of hot water and cave-ins. The accident rate in the deep mines led to one out of every thirty miners being disabled and one out of eighty killed. The small, unventilated work space was stifling. It was dust filled, and the candles and every exhalation added to the accumulation of carbon dioxide. The smells of decaying vegetable matter and human excretions added to the foulness.[31]

But Mackay was earning his stake, and in 1863 he had substantial interest in four comparatively unimportant mines: the Union, the new Oregon tunnel, the Caledonia tunnel, and the Milton. When the Milton was organized, Mackay became superintendent. Like all the principal companies, the group built an elaborate office near its mine, and Mackay lived in a room there. The mine was developed for several years, recovering little of its expenses, and later was absorbed into the Chollar mine.[32]

A newsman, Wells Drury, later speaker pro tem of the Nevada assembly, compiled a list illustrating the volatile nature of occurrences on the Comstock during the era. It included: sudden death in the mine depths; powder explosions; fires; blizzards; robberies; stabbings, street duels and barroom fights; rows in brothels; raids on Chinese opium joints; wild rumors of new strikes; the fevers of the stock speculators; gambling games for high stakes; and boxing, wrestling, and cockfighting. Another famous newsman of the era, Arthur McEwen, said that those who lived on the Comstock were "unbullied by authorities and indifferent to traditions. In their simple, sinful way, Comstockers enjoyed a life of audacious gaiety and gambling."[33]

Hundreds of professional gamblers had been attracted to the Comstock, and any saloon with a table could be the site of a game, but more urbane arenas were soon available. By the end of 1874 the *Virginia Evening Chronicle* announced: "Gambling here is carried on in a manner which removes many of its objectionable features." It touted the magnificent decor of the rooms, their five or six faro dealers being paid six dollars a day, and their banks, some as high as fifty thousand dollars. William Sharon was an inveterate poker player, often hosting high-stake games for mine owners and others of the upper economic crust.

Mackay allowed himself the diversion of poker until his wealth robbed him of its joy. A reporter told of Mackay, once he became rich, winning a high-stakes game at c Street's fashionable Washoe Club. "What of it?" said Mackay. "Even if I win every cent in sight it would not make the difference to me." He pushed away from the table.

" 'Leave me out boys,' [Mackay] said, 'I'm through! When I can't enjoy winning at poker there's no more fun in anything.' " Mackay told the others to keep his winnings and left. The others divided half among themselves and gave the other half to the county charity fund.[34] How often Mackay participated, even before it lost its appeal, is difficult to determine. He rose before dawn and worked into the night. Comstock historian Eliot Lord said he focused on "the whirl of the mighty lottery-wheel beneath his feet." He worked not to make money for money's sake, or to become a nabob, but "to win a name as master and manager of the greatest mines in the world—this was what he sought, as he declares."[35]

Early in his mining career, Mackay would impose on a friend, Adolph "Dolph" Hirschman, to interpret contracts for him. Although years later Mackay's penmanship was smooth and flowing, in the early days he was embarrassed by his cramped hand and would also ask Hirschman to write his letters. Mackay spent time reading and studying to overcome his handicaps. Other spare time was spent in the telegraph office learning to read Morse code messages. His habits were simple. Even after gaining great wealth, when he invited friends to share breakfast the fare did not vary: oatmeal, muttonchops, toast, and coffee.[36]

As he was gradually acquiring knowledge of the great silver lode, Mackay was garnering respect. Renowned banker D. O. Mills, one of Sharon's partners in the Bank Ring, remarked that from the beginning Mackay's "frankness of manner and his close application to business" assured his success. Grant Smith believed that his ability, uprightness, and steady judgment led to Mackay's association with the Bullion mine.[37]

The Bullion consisted of 1,424 feet between the Chollar-Potasi and the Gold Hill diggings. It covered The Divide, between Virginia City and Gold

Hill. Fifty-five co-owners, almost all residents of the area, held its stocks. Except for the Little Gold Hill mines, the Bullion was the only important mine not owned by San Franciscans. When San Francisco stockholders sold shares, other locals bought them.

J. M. Walker, whose brother served as governor of Virginia, incorporated the mine with a group of others in 1863. Walker, agreeable and well liked, became friends with Mackay and invited him to buy into the Bullion. It was an opportunity not to be missed, as the mine's prime location, between two leading producers, practically guaranteed that it held ore deposits. Mackay sold his other holdings, probably at great profit since there were plenty of buyers in what was a wild market, and bought in.

Walker was the superintendent of the Bullion, and Mackay became one of five trustees. Walker and Mackay soon became partners in all Walker's mining ventures. Apparently it was a verbal contract that each honored until they dissolved the partnership four years later. They conducted a profitable enterprise by building the Petaluma Mill at Gold Hill. The same cannot be said of their work in the Bullion. Assessing each of the 2,500 shares ten dollars every ninety days allowed the company to buy the most advanced machinery in the West. Walker was not a forceful personality, and in time he ceded the superintendence to Mackay, but even while Walker held the title, it is likely that Mackay ran the enterprise. The Bullion mine's shaft was sunk faster than any mine heretofore, and at each level drifts were dug two hundred feet wide. The mine was a will-o'-the-wisp: there was an abundance of low-grade quartz that held ore bodies, but none with enough value to be worth milling. Immense monies were spent searching, as signs consistently showed great promise. It was always believed that a few feet more was all that was needed.[38]

In 1864 Nevada mining prospects diminished, and the San Francisco Stock Exchange sagged. Wildcat mines found to be barren were deserted. In the large mines, expenditures were great and returns not as high as expected. The Ophir and Gould & Curry mines appeared to be played out. The leading mines in the boomtown of Aurora pinched out at the one-hundred-foot level, causing their stocks to crash and further drag down the Comstock market. The depression in Virginia City lasted into 1865. When the year began with storm after storm rolling from the sea across the Sierra into Nevada, the weather seemed to be mirroring mining prospects. It was during these hard times that Mackay won his first fortune in the Kentuck mine.[39]

The Kentuck was tiny, 94 feet, located in Gold Hill between the Yellow Jacket and Crown Point mines. The Yellow Jacket consisted of 957 feet, and the Crown Point, 541½feet. The Yellow Jacket had produced three million dollars' worth of ore beginning at the 180-foot level. But early in 1865, adding to the area's despair, there had been "a sudden impoverishment." The Kentuck had never produced pay ore, and the insignificant mine's value had long been depressed. In 1865 the Crown Point found its first body of ore at the 230-foot level. It was a west-dipping vein almost on the Kentuck border, making it likely that it extended into its neighbor. With their limited resources, Mackay and partner Walker immediately bought all available Kentuck stock. The Yellow Jacket then struck ore in a similar area on a course approximating the same west-dipping vein. If it was the same vein, the Kentuck would strike it too. Mackay took over management of the mine and quietly sunk a shaft using a horse whim (a capstan with radiating arms to which a horse is yoked) for a hoist.[40]

Where they acquired the money to buy into the Kentuck is uncertain. There are two stories, each illustrating a different Mackay trait. The more probable story said that they borrowed sixty thousand dollars from financier James Phelan, whose son later became the mayor of San Francisco and a U.S. senator. They paid the prevailing interest rate of 3 percent per month. When the loan came due, unable to pay, Mackay went to Phelan protesting vehemently that the interest rate was too high, demanding it be lowered. Phelan exploded, saying that the note would be renewed at 3 percent or not at all. Mackay grudgingly signed, then returned triumphantly to Walker, the ruse having gained them another three months.

George Lyman, whose father was a respected mine superintendent who worked for Mackay, published the second story, apparently taken from a Mackay friend. Because the event was said to occur in 1863 and the Kentuck was not deeded to Walker until two years later, it is certainly apocryphal, but the anecdote gives evidence of how some who knew Mackay perceived him. Lyman said that when the Kentuck owners decided to incorporate, one large shareholder was missing. The other owners offered a liberal bonus to anyone who could secure the missing shareholder's deed. Mackay disappeared during the summer and did not return until just before Christmas, when he produced the stock. The Civil War was raging, and it was said that he crossed through the rebel lines at Chattanooga to find the man fighting for the Union army.[41]

In any case, once Mackay and Walker became associated with the Kentuck, the small shaft was efficiently sunk. On January 1, 1866, at 275 feet, the Kentuck struck 10 feet of west-dipping ore, the same body as the adjoining mines. At the time the Yellow Jacket employed 180 men, raising 175 tons of ore a day. The Crown Point had 75 miners working and yielded 75 tons a day. The Kentuck was employing 11 miners, producing 10 tons of ore a day. Below 230 feet, where it was first struck, the vein bloomed, gaining from 40 to 65 feet at the 400-foot level, where it bottomed on a bed of clay. Over the next three years the three mines removed about three million dollars' worth from it.[42]

Walker was not destined for greatness. He later lost most of his fortune as a real-estate broker in San Francisco. In Walker's later years Mackay helped provide for him, and after his death Mackay assisted his widow. Years later Mackay commented, "When Walker had $600,000 he thought he had all the money in the world."[43]

Regardless of his financial acumen, Walker's geniality allowed a serendipitous consequence for Mackay. Walker was a friend to two other Kentuck stockholders: Irishmen who were to become key players with Mackay in the Comstock saga. James Clair Flood and William Shonessy O'Brien were partners in the Auction Lunch, a San Francisco bit saloon (where a drink cost a dime, as opposed to two-bit establishments, which charged a quarter and so were thought to be higher class). The bar was a few doors down from the Mining Exchange and, as in many saloons of the era, slabs of ham or corned beef were served free of charge. The proprietors added a touch of elegance to the concern by eschewing the traditional shirtsleeves uniform to dress in suits, with O'Brien greeting customers on the sidewalk in a high silk hat. O'Brien was amiable; Flood was astute, as well as being one of the best bartenders in the city.

Men of finance flocked to the place, as did the sports and Nevada miners, including Walker, who introduced the owners to Mackay. Men pursued a card game at one table, while important bargains were discussed at the next. The proprietors advanced money to miners, bought their claims, or entered into partnerships with them. These transactions, along with gossip and tips from customers, provided Flood and O'Brien with vast information about the mines. They soon opened a stock-brokerage office while continuing to run their saloon. By 1867, having earned sixty thousand dollars through investments, they would sell the bar to concentrate solely on the brokerage.[44]

Sometime in 1865 or 1866, Mackay met James G. Fair. Grant Smith said: "Socially [Fair] was as genial as the sun, his mouth dripped honey, and his tongue was smoother than oil, but his heart was a stone and duplicity second nature. On the Comstock he was known as 'Slippery Jimmy.'" He was also one the greatest miners of the era. His lifework and outside interests were the same: mining and mining machinery. His brother-in-law said that James Fair had no interest in books, music, or science but was totally absorbed in business.[45]

Fair, who was stocky, powerful, and athletic, was described by an associate as "one of the handsomest men I ever saw," with the "type of physique to attract attention anywhere." He had a native shrewdness and was indefatigable in attention to business. His affable swagger and jovial wit were appealing and could win the confidence of strangers. He was also a blowhard, and his propensity to exaggerate his own accomplishments while disparaging those of others, his cynicism, and his contempt for public opinion were decidedly unattractive. To associates he was crafty or wily; all others saw him as devious. Zinc Barnes, a jovial shyster himself, said of Fair: "The tears of widders and orphans is water on his wheel." And at his death, many years after his Comstock success, newsman Arthur McEwen commented: "Since James G. Fair died last week, his name has been on everyone's lips. I have yet to hear a good word spoken of him."[46]

Fair was a successful miner long before he arrived on the Comstock. He came to America from Dublin, Ireland, as a young teenager in the early 1840s. At sixteen Fair ran away from the family farm to work in a Chicago machine shop. A few years later he joined the gold rush, bringing his interest in mechanical devices west. In his mines he prided himself on using the finest hoisting works, pumps, blowers, and drills. In the 1890s he told an interviewer: "I learned as I went along. That sort of mining had never been done before. The methods I worked out are now being followed all over the world. Nobody has improved on them." Although Mackay and other Comstock superintendents and mining engineers should be afforded their share of the praise, there was truth in Fair's boast. In 1876, noting that the firm's Consolidated Virginia mine—with Fair as superintendent—was the most productive ever developed, the state mineralogist praised the systematic manner of the work, remarking that "all the improvements in mining machinery are found here."[47]

When Fair came to the West, he spent several years as a placer miner and several more as a quartz miner in the California Mother Lode country. He was sociable in those days, one of the boys. He became the superintendent of a stamp mill at Shaw's Flat in Tuolumne County, where he made some thirty to forty thousand dollars, but the profits were used up in litigation, and he moved to Angels Camp. There he owned interests in at least two mines. He directed the mining and attended to all details of his mines, did the blacksmithing and the carpentry, and took part in the pick and shovel work. An intimate who worked with him said: "Fair was the handiest all-around mining man I ever saw, and I do not believe there was ever a more thorough one on this Coast or anywhere."[48]

Fair met Theresa Rooney, a pretty and amiable widow and the daughter of a rancher, at a dance at Angels Camp. She was the belle of the territory, "a comely young woman who was admired of all the young men who passed that way." They married, and the first of their four children was born in Calavaras. In 1865 he disposed of his California mine holdings and they moved to Virginia City.

"When I went to Virginia City," Fair said, "all I wanted was to get a little bit of a stake to keep my family from starving." Shortly after his arrival, owing to his wide range of experience, he was appointed superintendent of the Ophir mine. After a brief stint there, he became associated with the Hale & Norcross mine, apparently serving as a foreman or assistant superintendent. He later claimed to have been the superintendent, but mine records list C. C. Thomas, who later served as superintendent of the Sutro tunnel, in that position. Fair was discharged on November 28, 1867. One account maintains that he was dismissed for insubordination; one of Fair's versions of the story was that after he put the mine "on a paying basis," a relative of the owner replaced him. His brother-in-law later said he got the story "from the inside" that Fair had stepped aside as a ruse to trick Bank Ring magnate William Sharon, the Hale & Norcross owner, so Fair and his partners could get control of the mine's stock. Fair's tendency toward self-aggrandizement and prevarication, and the fact that he did not resign but lost the position, make this last story decidedly less plausible. In any case, in spring 1868 he moved to southern Idaho.[49]

In 1867 Walker sold his interest in the Kentuck to Sharon for a fortune—six hundred thousand dollars. At least one Bank Ring member, Alvinza Hayward, was also a stockholder, so Sharon now controlled the mine.

Mark Twain, living in Washington, D.C., in December 1867, noted that

J. M. Walker had appeared there occasionally, that he was speculating in lands in Virginia, and that he bought a home at Binghamton, New York, for twenty-five thousand dollars. In 1868 Walker served as a Nevada delegate to the Republican Convention, which nominated Ulysses S. Grant for president. It was widely reported that Walker then left to travel in Europe. After assisting Mackay in his start to riches, he never returned to the Comstock, but his affection for his old partners is apparent in the naming of three of his children: James Flood Walker, Mary Flood Walker, and John Mackay Walker.[50]

At the end of November an associate of Alvinza Hayward, J. P. Jones, arrived on the Comstock after losing the race for California lieutenant governor. He was immediately engaged as superintendent of the Kentuck. Mackay continued exploring the Bullion mine's depths for two years. The Bullion's owners had spent a million dollars, and Mackay sank the shaft to 1,400 feet before they agreed that it held no ore and they sold out.[51]

During this time, Flood and O'Brien opened their brokerage in an upstairs office on Montgomery Street in San Francisco. The prince of San Francisco stockbrokers, James Keene, said Flood had more natural talent for stock operations than anyone he had ever known. Flood's ability was about to undergo its first major test.

When Fair returned to the Comstock late in 1868, Mackay enlisted him. They would go in with Flood and O'Brien on a business enterprise. Walker was to have been included but decided not to participate. The partnership was divided into eighths, with Mackay using his mine profits to gain ⅜ and become majority shareholder. Flood and O'Brien borrowed in excess of fifty thousand dollars and split ⅜. (The lender was Bank Ring member Edward Barron, who later joined Mackay and the others in further undertakings.) Flood and O'Brien assisted Fair with his ⅖ share. Fair gave them promissory notes secured by his future employment as superintendent and potential profits.[52]

In their mining ventures Mackay acted as the administrative head, Fair as the general superintendent, and Flood as the financial representative managing stocks. Which of the members was the true leader was often a topic of speculation. Comstock broker George Mayre Jr. said: "The question was never answered, for the reason that there was no general leader, each of the three taking the lead in those activities which were tacitly recognized as falling more particularly within his field." There is no recorded comment by

Mackay addressing the topic. Fair, called by Grant Smith "the best of self-advertisers," continually bragged that it was he who served as leader, referring to the others as "lads" and, when they fell in his disfavor, "kindergarteners." Flood, intimating it was otherwise, commented: "I know only one man in the world who can break me, and that is Mackay."[53]

The genial O'Brien was included in the partnership because of his friendship with Flood, which lasted until O'Brien's death in 1878. Likewise, Mackay and Flood developed an enduring friendship. The best to be said for Fair's affiliation was that he brought an expertise to the association. His original camaraderie with Mackay deteriorated over time into uneasy tolerance on both sides, then a falling out, and finally, a parting of the ways. The alliance, which was oddly effective for several years, immediately undertook a seemingly impossible task. They moved to wrest control of the Hale & Norcross mine from the absolute monopoly of William Sharon and the Bank Ring.[54]

CHAPTER 3

Hale & Norcross

MARIE LOUISE ANTOINETTE HUNGERFORD BRYANT, called Louise, had a story unique to the West in the era. She was the daughter of Daniel E. Hungerford, who had served as a captain under Winfield Scott in the Mexican-American War. After living in New York City, Captain Hungerford returned to the West during the gold rush. In 1854 his wife, his mother, and his daughter Louise followed him to Downieville, California. In 1857 a second daughter, Ada, was born to the Hungerfords. She was fourteen years Louise's junior, and the parents sent Louise to St. Catherine's Female Academy, in Benicia, California. The Sisters of St. Dominic, some from Mexico and France, refined Louise's knowledge of Spanish and French. They also taught embroidery and the scholastics previously taught by her mother and grandmother and introduced her to playing piano. The following year, with her parents unable to further finance her education, Louise returned to Downieville. She fell in love with twenty-four-year-old physician

Edmund Bryant, a nephew of William Cullen Bryant, and married Edmund as soon as she turned sixteen.[1]

Hungerford, serving as a major in the Sierra Guards militia, survived the routing of whites in the Paiute Lake War of 1860. In spring 1862 he won praise commanding a regiment at Fair Oaks, Malvern Hill, fighting for the Union in the Civil War. He returned to California with a scheme to raise a regiment and attack the rebels through Texas. But the orders from Washington never came and, with his wife and younger daughter, he moved to Virginia City. In March 1863 Dr. Bryant followed his father-in-law to the Comstock, and that summer Louise and their daughter, Eva, joined him.[2]

The young Bryant family had problems in Virginia City. A second daughter, months old, died of smallpox. Opium and its derivatives were not uncommon on the Comstock in the 1860s and not outlawed until 1876. Bryant began drinking heavily and using morphine.[3] Left in her father's charge, one afternoon little Eva fell down a flight of stairs, breaking her hip. She required a long recuperation period and walked with a limp the rest of her life. In 1864 Daniel Hungerford moved to San Francisco and Bryant left his young family for other goldfields. Louise, Eva, Mrs. Hungerford, and Louise's sister, Ada, remained in Virginia City. In spring 1866 Bryant fell ill and died at Poverty Hill, near La Port, California.[4]

For some time Louise Bryant had been sewing for the ladies in town to earn an income. Her hand embroidery was so fine that it was always in demand. The legendary Comstock priest Father Patrick Manogue thought to help her further, recommending that Sister Frederica S. S. Xavier employ Louise to tutor French and piano at St. Mary's Catholic School. Even with the additional work, Louise's income was meager for her circumstances, but she was young and attractive, and six months after Dr. Bryant's death, suitors began to appear. Two were Mackay's friends. Dolph Hirschman was attentive to her and said to be very kind to the family. Years later Hirschman's son said that "Louise was a very beautiful girl and woman, with brown hair and wonderful blue eyes," and he had a different view of the relationship, stating that his father interceded for Mackay with the widow. Harry Rosener, who owned a leading mercantile in town, was also kind to Louise. Mrs. Robert Howland, Louise's friend dating back to 1854, who lived close to her in Downieville and Virginia City, said: "It was thought for a time that she might marry Harry."[5]

The story has been widely spread that Mackay and others, hearing of her financial plight, started a subscription, and that Mackay and Rosener carried the money to the young widow. But Mrs. Howland said that Mackay would never have carried subscription paper around to raise money: "Mr. Mackay, so modest and retiring, would have given the last cent he had to have alleviated distress, but [he was] too proud spirited to have sought help from others, and hated publicity as a good Samaritan."[6]

Theresa Fair was a genial spirit, devoutly Catholic and hospitable. On Christmas day 1866 she invited Louise and daughter Eva to dinner. She also invited Mackay, who at the time shared ownership with Fair in a mine, presumably the Occidental. Much of Mackay's courtship was to take place at the Fairs' house. The affable couple encouraged the relationship between the grave and serious thirty-five-year-old Mackay and the somewhat reticent twenty-four-year-old widow. Mrs. William Mooney, who lived between the Fairs and Mrs. Bryant on A Street, said that Mrs. Bryant and Mrs. Hungerford "were having a hard time to make ends meet" in 1867. Mrs. Fair often invited Mrs. Bryant to dinner, and Mrs. Mooney would host her mother. Louise Bryant, who had been in love with Dr. Bryant, had been in mourning only half a year when she was brought together with Mackay, ensuring a civil period of courtship. Another factor slowed the romance: she was not certain that Mackay was right for her.[7]

Mrs. Howland commented that Mackay was "well-poised and reserved in manner" but that the widow Bryant's ambitions were "social." Howland reported that Mrs. Bryant was "reluctant to marry an uneducated though successful miner, who stuttered somewhat; mindful as she was of the cultured though tragically unfortunate gentleman who had been her husband. Her intense ambition to shine had not been dulled by years of deprivation."[8]

When Mackay won the widow's hand eleven months after their first meeting, those who knew them were delighted. It was a small, private ceremony, but he sent a case of champagne with the announcement to each of the territory's three newspapers. On November 27, 1867, the *Territorial Enterprise* stated:

Married
In Virginia, November 25, by the Rev. Father Manogue, J. W. Mackay to Louisa Bryant. (No Cards)
It is seldom that as brief an announcement affords us so much gratification, or that a case of Krug honors the chronicling of as happy an event. The union of

so estimable a couple and the devotion of a thousand worthier friends make every wish of joy and prosperity which we could utter superfluous; and so we simply offer the congratulations which all who know them must extend to two so worthily mated that none can say which made the better choice.

Mackay soon purchased a lot for five thousand dollars at Howard and Taylor Streets, high above the business district in Virginia City, and, with his profits from the Kentuck mine, built a house. With its furnishings of brocade, polished oak, and carpet from Turkey, it was worth perhaps twenty-five thousand dollars. The Mackays, Mrs. Hungerford, and Ada moved in.

Mary McNair Mathews, a seamstress on the Comstock, told of leaving the employ of Mrs. Mackay's mother in 1868 shortly after being hired. The Mackays were doing well financially but were not yet wealthy. Rather than the going rate of three dollars a day, Mrs. Hungerford was paying Mathews one dollar and providing room and board. But the seamstress was relegated to taking her meals with her son in the kitchen, being served when the family was through with each course in the dining room. Not believing she could raise her son properly eating at the second table, Mathews protested to Mrs. Hungerford.

Seeing that the seamstress's feelings were hurt, Louise's mother explained, "Mr. Mackay would never eat with hired help." Mathews said that she could not go on under the present circumstances. Mrs. Hungerford replied, "Well, I am not to blame, you know, but he is very particular, and if you had rather board yourself, I will pay you in provisions, if that will do." (Mathews later commented, "I suppose Mr. Mackay was more particular now than when he was a common miner, working for his $4 a day and *packing his dinner-bucket.*") A time later, living in her own place, Mathews sent word to Mrs. Hungerford that she could not come to work, because her son was very sick. Mrs. Hungerford arranged to have milk sent for the sick child. Mathews said, "Sometimes she sent me a hot loaf of bread, and a loaf cake," and added that she was very kind throughout her son's illness.[9]

Louise Mackay's mother was described by Alexander O'Grady as "a motherly soul, always anxious to be helpful to everyone, both within and without the house. When any of the servants were ill she attended them with as much care as if they belonged to her own family." But Mrs. Hungerford's contention that her son-in-law did not wish to eat with the help was almost certainly untrue. Nowhere else in all that was written about John Mackay is there mention of him wanting to affect social airs at meals. As Mathews back-handedly noted, he was used to eating with miners, and it was often reported that he preferred simple fare with friends or associates to elegant meals.[10]

Although only the year before Louise had been a seamstress herself, it is much more likely that Mrs. Hungerford or Louise did not want Mathews and her son dining with the family. Mother and daughter were persistent in their attempts to gain social status. All who have written about Louise mention her social ambitions. With only one year at St. Catherine's Academy, she was on her way to making herself into a cultured and refined lady, with servants and a footman. The mother used her son-in-law's wealth to affect a similar lifestyle. It appears, in this instance, that the ladies were beginning their social climb on the Comstock at the seamstress's expense.[11]

Shortly before the Mackay-Bryant wedding, William Sharon had purchased controlling interest in the Kentuck mine. Sharon was the most powerful man in all Nevada. Before a year passed, Mackay began the struggle against him that would lead to control of the entire Comstock.

William Sharon came to the Comstock at the beginning of the mining depression in 1864. He represented the single most powerful entity in the West, the Bank of California. Sharon opened a branch office for the bank in Virginia City on September 6, 1864, setting up a regular payday for each of the mines and loaning money at, for that time and place, the reasonable rate of 2 percent a month, 24 percent a year. At a time when others believed the Comstock Lode was played out and were abandoning the territory, Sharon persuaded his partner William Ralston to invest heavily, using private as well as bank funds.

Too many mills had been built in the rush to make money on the Comstock. Those working rich mines built them, as did those expecting to find riches. Others were built as custom mills. What seemed a promising business quickly became burdensome to sustain. Mill owners were unable to earn enough to pay their debt to the bank, and Sharon foreclosed. By June of 1867 Sharon and Ralston took possession of seven mills and formed the Union Mill and Mining Company. They and their associates, known as "The Bank Ring," bought into mines so that boards of trustees were appointed to serve at their pleasure. Sharon directed the mines to use only Bank Ring mills, squeezing out formerly productive mills. Within two years Sharon's Ring owned seventeen mills and controlled all the important mines.[12]

Sharon's machinations insured that his properties turned profits. The most glaring example was the Yellow Jacket mine. It worked rich ore from 1863 to 1872, but its assessments in that time nearly equaled dividends paid. In a typical maneuver, Sharon ordered that all Yellow Jacket ore be processed at a Ring mill, letting its own mill stand idle. In a particularly egregious expansion of the scheme, he sold his Yellow Jacket stock and directed the mine superintendent to mix waste rock with the pay ore. This cost the mine tens of thousands of dollars extra in milling expense, while enabling the Bank Ring mill to run twenty-four hours a day. Mine stockholders were cheated out of dividends, while Sharon and his handful of partners in the Union Mill and Mining Company reaped enormous profits. Years afterward the *Virginia Evening Chronicle* sarcastically reported: "Mr. Sharon being controlling trustee of the Yellow Jacket mine, sent to the mills of the former company barren 'Jacket' rock by the thousands of tons, and generously permitted the mills to crush it at $12 a ton. . . . This sum the Jacket stockholders had to pay by assessment."[13]

Sharon and his Ring created a vertical monopoly that ruled the territory. He paid mining engineers to inspect all the mines on the Lode and give him daily reports on the veins that were widening and those that were diminishing. He used the information to manipulate the stock market, buying and selling with each fluctuation.

Sharon controlled the timber and lumber operations. He won a decision against Carson River farmers and ranchers in the Nevada Supreme Court. The high court, reversing lower-court rulings, restricted diversions of the river for irrigation, saving it to power the mines' mills located downstream from the agricultural lands.[14] He took control of the area's transportation by appropriating money pledged to Adolph Sutro's tunnel project and building the Virginia & Truckee Railroad. He even bought the local water company, ensuring that in dry years serviceable water was provided to his myriad operations, if to no one else.

The courts supported Sharon, and he directed a majority of legislators. Taxes on ore from his Storey County mines were periodically reduced, until in 1867 the law allowed manipulation to the effect that in a ton of ore only one dollar of profit might be assessed. His railroad was financed almost exclusively by county bonds, which he pledged would be repaid through substantial taxes once the road was built. The promised taxes were never collected. H. H. Bancroft's history notes: "So far from growing any richer through the possession of a railroad, which was making $12,000 a day, the total tax paid to the county by the company in twelve years was very little more than the interest the county had to pay to the company on its bonds. . . . That Nevada assessors, sheriffs, legislators, and shareholders have assisted these railroads to oppress the commonwealth cannot be gainsaid."[15]

In 1870, a newspaper, *The People's Tribune,* was begun in opposition to Sharon's interests. Its initial issue editorialized: "It has long since become a fixed opinion in this community, that under the almost autocratic sway of Wm. Sharon, the business of a number of our mines . . . if done in a similar way in any well governed State of Europe, not only Wm. Sharon but (with only an exception or two) his subservient mining superintendents, secretaries and foremen in the mines, would be promptly consigned to the public galleys or be locked behind the iron gratings of prisoners' cells."[16] The day the paper was published the editor was accosted and beaten on the street, and days later he was confronted and beaten again by Sharon's Yellow Jacket mine superintendent.

A state legislator, F. E. Fisk, reported that he knew other members of Nevada's Ninth Legislature were "bought, body and soul, and money was paid almost openly" by Sharon's railroad interests. Fisk was forced from office because of his opposition to the Sharon machine: "It was intimated that I would be placed in a position to make money by not taking an active part in opposing them, and after trying all means they attempted to bluff me, and threatened to injure me in business, etc., and misrepresented me in every way, and put up jobs to get me into trouble." Sam Davis, a newsman at the time, observed: "Everybody was at the mercy of the unscrupulous combine."[17]

Sharon and his Bank Ring, attempting to acquire all the paying mines, in 1867 turned their attention to the Hale & Norcross mine, located between the rich Savage mine and the equally prosperous Chollar-Potasi. Its owners had originally spent $350,000 digging a deep shaft without finding ore. They persisted and, in December 1865, made a rich strike at the six-hundred-foot level. During the next two years, the Hale & Norcross produced $2,200,000.[18]

There was a familiar practice on the Comstock, utilized at times since 1863. When a mine was being developed at a new level, as it reached the Comstock Lode's ore-bearing east wall, miners would be confined to the mine and no one else was allowed to visit. This was done so that information about what was found that would affect the stock market was kept for the exclusive use of the directors. In January 1868 owner C. L. Low and super-intendent C. C. Thomas used the device when piercing the wall in the Hale & Norcross at the 930-foot level. They confined twenty-five miners in the hoisting works for three days. The workers were paid twelve dollars a day, three times their usual wage, and so were willing captives.[19]

When the crew was released, word spread that rich ore had been found. Sharon had already drawn a bead. The Hale & Norcross election of trust-ees was two months away, and a bitter bidding war broke out for control of the mine's voting shares. The mine's eight hundred shares went from $300 apiece on January 8, 1868, to $2,200 per share on January 11, the day the miners were released, to $4,100 on February 11. Two shares sold for $10,000 each on February 12. Sharon's group, with seemingly limitless resources, ac-cumulated enough stock to defeat Low and his friends, and Sharon's board of directors was seated on March 10, 1868.

The rich strike proved ephemeral, and the mine produced a trifling sum in 1868. Sharon, the first to be apprised of the weakness of the find, used a favorite trick. He sold his stock, and his board levied an assessment. Two other assessments followed, bringing the year's total paid by shareholders to $201,960. Sharon's stratagem of selling his shares before issuing assess-ments was effective, because his superintendent ran the mine. Sharon would be the first to know when the property was about to again improve, and he would buy back shares at bargain prices before any dividends were paid. That year the board increased the number of shares from eight hundred to eight thousand, and, with the mine still unproductive, the value of a share fell to $41.50.[20]

At that time James Fair returned to Virginia City, ending his self-imposed exile. Having worked in Hale & Norcross, he knew the upper levels had not been mined efficiently—ore remained that if removed systematically would sustain further development. He also believed its immense unex-plored lower territories might well contain ore. He and Mackay discussed the mine's possibilities.

Although early in his career, as Mackay moved from one mine to another, he may have seemed to be gambling, he took few risks in his maneuvering.[21] Mine superintendents were allowed to freely inspect each other's mines, and it can be assumed that Mackay had been inside and studied the Hale & Norcross. There were two things he had to decide: whether he agreed with

Fair about the mine's potential, and—being married only a few months, with a wife, stepdaughter, niece, and in-laws to support—whether he was willing to hazard a gamble. He decided in the affirmative to both, contacting his associates in the Kentuck, Flood and O'Brien. The Mackay, Fair, Flood, and O'Brien partnership was formed with the foolhardy purpose of assaulting Sharon and the Bank Ring.

In San Francisco Flood quietly began buying Hale & Norcross stocks. In order to keep his transactions secret, he did not mark the purchases on the books but merely locked the certificates in a tin box in the safe of the well-established Parrot's Bank. January passed, and neither Sharon nor the other brokers on the exchange were aware that anything out of the ordinary was taking place. When the subterfuge was finally uncovered, Sharon's agents scrambled to buy. Prices of shares again rose to spectacular heights.

An oft-repeated tale about the battle for ownership may be only a legend. Still, it has elements of truth in it and is too interesting to pass over. It begins with the approaching election of the Hale & Norcross board, the voting shares split evenly between the two sides. No stockholder was willing to sell at any price. Finally, an operative sent word to Sharon in Virginia City: an elderly widow living near San Francisco owned one hundred shares and was unaware of the great bidding war. Sharon hurried to the telegraph office to wire William Ralston, his partner in San Francisco, using a secret code to tell Ralston to buy out the widow's shares.

When Sharon left the telegraph office the operator, who resented Sharon and the monopoly, broke Sharon's code, deciphered the message, and gave it to Mackay. Mackay sent his own cipher to his partners in San Francisco. The next morning Ralston sent his man to purchase the shares only to learn that Mackay's partners had been there the night before and had paid the woman a small fortune for them.[22]

Without mention of the dramatic widow incident, on Friday February 26, 1869, twelve days before the mine's election of officers, the *Territorial Enterprise* casually announced in its "Mining Intelligence" column: "J. G. Fair and John W. Mackay of this city now own a controlling interest in the Hale & Norcross mine. The mine is looking exceedingly well, and the gentlemen named will doubtless shortly become millionaires."

On the 27th the *Gold Hill News* reported that a five- or six-foot vein of good ore had been found between the fourth and fifth levels of the Hale &

Norcross. The report continued that since Mackay and Fair owned over four thousand shares of the stock, they would be able to control the choice of officers for the ensuing year.[23]

It was a bitter blow to Sharon to have good ore discovered just as he lost control of the mine. Thomas H. Rooney, Fair's brother-in-law, later said that a week before the election, Fair took up the mortgage that Sharon held on the Bacon, a twenty-stamp mill in Silver City. Rooney said that Fair told Sharon he thought he would "work a little Hale & Norcross ore" in the mill, giving Sharon the "first inkling" that the property had passed from his control. The announcements having already been published at the time of the mill's purchase, this seems merely another instance of Fair's passing along an anecdote for the purpose of self-aggrandizement.[24]

The new partnership's well-executed plan almost collapsed in the end. Flood did not know that stock had to be transferred on the book before it could be voted. He had the stocks in the box at Parrot's Bank but still had not entered his transactions three days before the annual meeting. Luckily, a man named "Uncle Billy" Watson mentioned to him that if stocks were not transferred, the men in whose name they were issued held their voting rights. Flood hurriedly obtained one certificate for the entire amount.[25]

On March 10, 1869, ownership certificate in hand, Flood controlled the election. It resulted in the Ring's trustees being removed and Mackay and O'Brien's being selected as board members. Flood was elected president of the company, and Fair became the mine's superintendent.[26]

The defeat of Sharon was the first crack in the bulwark of the Bank Ring cartel. Shortly thereafter, the Mackay group began to crush the Hale & Norcross ore in its own mills, creating another fissure in the monopoly. Mining stockbrokers and the general public now took an interest in the new firm. Broker George Mayre Jr. said that many inquiries were made about them without turning up anything of particular significance, but that "what they had recently done and especially the manner of its accomplishment, gave promise of much interest in their future."[27]

The group, later known as the Bonanza Firm or simply the Firm, took immediate action after gaining control of the mine. The difference between their administration and that of Sharon was apparent. Before the management change, the Sharon group had ordered another eighty-thousand-dollar assessment. Mackay rescinded the order and returned money already paid.

As soon as Fair verified that they were working in good ore, a dividend was declared.

Commenting on Mackay's character, Grant Smith said: "Men always knew where he stood. So truthful that even cynical old John Kelly [a stock market bear notorious for shorting Hale & Norcross stocks] admitted, 'John Mackay never lies.'" Years later Flood's chief clerk said that Flood routinely shrugged off personal attacks against him, but "let anyone so much as hint that the firm was withholding a disproportionate share of the bonanza profits and his cries of rage could be heard in the next county."[28]

Mackay, as general manager of the Firm's affairs, was confident and decisive. Experienced in every facet of mining operations, he joined Fair in managing the mine, although Fair held the superintendent's title and received the salary. Mackay quickly learned about milling and supervised that process as well. He was attentive to detail, staying on task in his busy schedule but often beginning a new enterprise before completing the last.[29]

Fair's abilities were given full rein in the new undertaking, and he exhibited his unique fitness for the role. The newest machinery was ordered, and he adapted it to the needs of the mine. He studied the tunnels thoroughly, his sharp eye and acute judgment seeming instinctive. Fair possessed unflagging energy, appearing in the mine at all hours of the day and night. Journalist DeQuille said: "No one knows exactly what 'Uncle Jimmy,' as the 'boys' call him, is up to. You see the hole by which he goes into the ground, but when once he is down out of sight you never know in what direction he is drifting." He ran the mine in autocratic fashion, causing Eliot Lord to comment: "His watchfulness was alleged to approach espionage, and his devices for detecting breaches of duty were sometimes more apt than commendable."[30]

Several months after taking ownership of the Hale & Norcross, Mackay and Fair ran afoul of some stockholders and the *Gold Hill News*. The *News* was the leading mining journal in the state. It was not afraid to take a side in political frays, and often its support reflected a good relationship with Sharon and the Bank Ring. Philip Lynch was the paper's owner and editor. Toward the end of 1867 popular newsman Alf Doten became the paper's associate editor, sharing the writing duties with Lynch (upon Lynch's death in February 1872, Doten purchased the paper).[31]

On July 10, 1869, the *News* announced that the good body of ore at the 1,030-foot level of the Hale & Norcross showed no signs of failing. "This same body of ore will doubtless be found to extend at least 100 feet deeper. The old upper workings of the mine still also yield considerable, very fine ore. The dividend of $6 per share, declared last Tuesday and payable to day [*sic*], amounts to $48,000 in the aggregate." That month Mackay used Hale & Norcross proceeds to purchase the French and Sullivan Mills. The works, which adjoined each other, were reconfigured and enlarged to assist in crushing the pay ore. The Sullivan Mill had not been used in four years, so the rebuilding was extensive, Mackay stripping out and replacing its old machinery.

But the local economy again began faltering. Lesser mines shut down, while others were operating with reduced work forces. On October 4, 1869, unnamed Hale & Norcross stockholders called a meeting in Virginia City and another in San Francisco to investigate company accounts. In an editorial the *News* remarked that while they opposed that mode of warfare on mining companies, they felt compelled to aid the distressed stockholders.

The week before, on September 29, 1869, when an advertisement in the *Territorial Enterprise* attacked Sharon's management of the Yellow Jacket and requested a meeting of all who were dissatisfied, the *News* had derided the notice. It pointed out that the announcement was published anonymously and that officers had been duly elected only two months earlier. "Every stockholder in that company had a fair chance at the election, and he could vote, according to his interest, just for whoever he pleased."[32] Now, days later, regarding the Hale & Norcross, the paper editorialized:

> We have given but little attention to this mine, but we have noticed a load of Hale & Norcross bullion pass our office almost daily, and heard reports that the mine was looking well, and producing a large amount of high grade ore. We are inclined to believe these reports, judging from the large amount of money invested in mills by the present management. The mine is evidently paying *somebody,* and we are not surprised that a meeting of stockholders has been called to inquire into the affairs of the company. The controlling powers of our mines must not forget that at law "the rights of the weak and the strong are equal."

The stockholders' inquiry came to naught, as the purchase of the mills enhanced the company's efficiency and, once they were operating, the mine paid another dividend. At a time when most Comstock mines were levying assessments, the Hale & Norcross paid $192,000 in dividends for the year.

They produced the profits despite the fact that their ore milled at only $25 a ton—well below some nonpaying veins of ore. Exceptional management had allowed them to survive the first of many public attacks. In 1870 Fair ordered a crosscut sixty-three feet into what was believed to be the Lode's east wall and struck a rich new body. Dividends rose to $536,000.[33]

In an incident that was said to occur once the Hale & Norcross was seen to be profitable, a man who had a fifty-thousand-dollar investment in the Yellow Jacket mine proposed a scheme to Mackay to gain control of it. The Yellow Jacket was mining rich ore but nefariously milling waste rock with it, and it was paying no dividends. Although the five Yellow Jacket trustees were Sharon appointees, two of them apparently had had enough; a third trustee could be bought for twenty-five thousand dollars. They agreed with the investor that Sharon's superintendent should be replaced, and they wished to appoint Mackay the new superintendent. But when approached, Mackay rejected the idea, saying he did not want mining to become a cut-throat game. "If we should get control of Jacket that way, we never could be sure of our own mines, for Ralston and Sharon would be playing the trick back on us."[34]

As his mining prospects continued to improve, Mackay's family life became more complicated. Louise became pregnant and, as her due date approached, went to stay at the Grand Hotel in San Francisco. She gave birth to John William Mackay Jr. on August 12, 1870. She never again was satisfied with staying on the Comstock. On Christmas day that year Mackay adopted Eva and insisted that they take her to France, where doctors might be able to correct the severe limp she had suffered since breaking her hip. Shortly thereafter, John, Louise, Eva, John Jr., Louise's parents, and her sister, Ada, traveled to Paris. John visited Dublin and London as well as Paris then returned to the Comstock alone. For two years the extended family remained in France. Eva and Ada attended French schools, and Eva underwent a surgery that was a partial success, her limp becoming less pronounced. The family, tapping into John's fortune, spent the winter on the Riviera and lived the rest of the year in Paris.[35]

CHAPTER 4

Competitors

IN THE EARLY 1840s, when young John Mackay was hawking newspapers on the New York streets, newsman James Gordon Bennett influenced him. Bennett began publishing a one-cent daily, the *New York Herald,* for working-class readers in 1835. Of his own work ethic, Mackay later said: "I went West with the high tide of the gold-seekers, and roughed it with the rest, in the sole ambition to make myself an equal to the hero of my boyhood . . . James Gordon Bennett."

Bennett was vigorous, brusque, straightforward, and relentless. Society loathed him; the lower classes were his public. Bennett was the first even-handed Wall Street reporter, unafraid to write up its swindles and name names. He played up scandals in an era when such things had traditionally been reserved for parlor gossip. Bennett was described as an "ill-looking, squinting man," to which he replied that at least he was unlike his antagonists who were "squint hearted."[1] He was a single-minded fireball, tweaking the dominant figures of the day in support of the underdog. Mackay adopted some of Bennett's competitive traits, and they came into play in his association with three larger-than-life Comstock characters: William Sharon, Adolf Sutro, and James Fair.

In August 1869 U.S. vice president Schuyler Colfax and an entourage that included renowned eastern newspaperman Samuel Bowles arrived to visit the mining district. Nevada senator James Nye accompanied them to Virginia City, where Mackay, Fair, and their wives were among the welcoming party. That night Sharon hosted a reception at his Victorian home, including his rivals on the guest list. A brass band played, colored lanterns hung outside, and a crowd of onlookers filled the street. Flowers filled the interior, lending a gala air to the evening. Louise Mackay commented that the affair allowed her to mingle with the social elite for the first time. It was an event that foreshadowed her life in distant, fashionable cities.[2]

An incident some months later reveals something of the financial relationship and bonds that linked the opposing mining bosses. Even with the Hale & Norcross working in good ore, Mackay had not remained idle. He invested in an Idaho mine, losing a considerable amount of money (three hundred thousand dollars, according to one source) and jeopardizing a Nevada stock transaction. Mackay ran into Billie Wood, one of the Bank of California's attorneys, and told him: "I must have $60,000 today or lose stocks which in three months would make me twenty times $60,000." Wood took Mackay to his office above the bank and explained the situation to his partner, Thomas Sunderland, a principal bank stockholder. Sunderland made a memorandum of the stock and left the office. Returning a few minutes later, he had Mackay sign a note and gave him Sharon's personal check for the $60,000. Sunderland told Mackay that Sharon was glad to loan him the money, and Sunderland offered some advice. Commenting that both Mackay and Sharon were hot tempered, he said, "When you both feel like fighting at the same time, separate and fight outsiders."[3]

In the zero-sum game of economic survival of the fittest, Sharon and Mackay were in direct competition: riches taken by one reduced those available for the other. Because of the expense involved in the process, quartz mining was necessarily a game for those with large amounts of capital. The stakes were raised considerably when Sharon formed the monopolizing Bank Ring, using money from some of the wealthiest men in the West as well as the seemingly limitless money of the Bank of California (three million dollars of bank money was invested in the Comstock at one time). Still, as evinced by Sunderland's statement, there was a fraternal order that involved only the Comstock's prominent players. As they fought against elements that only they confronted, others were seen as distinctly foreign "outsiders." Hence, as a courtesy, all superintendents were allowed access to others' mines, and lesser partnerships and side agreements between competing entities were fairly common. The skein of Comstock mining relationships is suggested by Sharon's $60,000 loan as well as, a short time later, his sale to Mackay of ¼ interest in the Petaluma Mill in Gold Hill; Fair and Sharon later co-owned the Stewart, Kirpatrick & Company quartz mill. When Mackay and Fair struck their bonanza in 1872, they gained access to it through the Gould & Curry, an adjoining mine owned by Sharon.[4]

Adolph Sutro was another remarkable character whose work affected Mackay's interests. Sutro had a plan to remove the volumes of water that flooded the Comstock mines far underground. He proposed digging a great tunnel beneath the mines for use as a drain. At the same time it would serve as a transportation corridor, efficiently moving men, materials, and ore. In April 1866 twenty-three mining companies, representing 95 percent of the mines of value on the Lode, subscribed to Sutro's tunnel project. They agreed that when the tunnel was connected to their mines, they would pay a fee of two dollars a ton on pay ore shipped through it as well as a stipend for using it to move work crews, waste rock, and supplies.

On January 15, 1868, William Sharon and eight influential mining super-intendents signed a one-sentence telegram to the Nevada senators in the U.S. Congress reversing their support of Sutro: "We are opposed to the Sutro Tunnel project and desire it defeated if possible." Mackay signed as the superintendent of the Bullion Company.[5] The other mining companies eventually followed Sharon's lead, and all support for Sutro, including six hundred thousand dollars in pledged subscriptions, was withdrawn. Motives for the disavowal of support remain murky 150 years later.

When he first arrived on the Comstock in spring 1860, Sutro proposed that mining should be done from a tunnel begun low on the hill, then the mining shafts sunk to meet it, at once insuring drainage and ventilation.[6] Instead, all mines began on the surface and sunk their shafts. When various mines encountered vast amounts of water, they began drainage adits like the one Mackay built for the Mexican in 1860. Because of costs and difficulties in construction, the largest-scale drainage drifts were abandoned before completion. In 1864, when many disputed claims were being adjudicated and compromises were struck so mining could proceed, Sutro proposed to build the all-purpose tunnel that would allow more efficient development. Sutro's pledge to raise three million dollars for the tunnel from eastern capitalists before the end of August 1867 made the project even more attractive. By April 1866 the mine companies' executives signed on.[7]

During the following year, Sutro's egotism and aggressiveness put off many leading Comstockers. The *Territorial Enterprise* noted Sutro had declared that upon the tunnel's completion his new town, at the tunnel's mouth on the Carson River, would cause Virginia City and Gold Hill to be depopulated—their buildings sold for the price of the nails holding them together. The idea panicked Comstock business owners.[8]

A further problem was that Sutro's tunnel had charters from the state legislature and the U.S. Congress, and people were suspicious of corruption in deals involving government charters. In the middle decades of the nineteenth century, corruption in business became flagrant. During the Civil War years, government spending caused a universal scramble for assistance. Councils, legislatures, and the U.S. Congress were bribed into awarding franchises and charters for railroads, public utility systems, and other projects. The public, including workers and the capitalists' victims, saw that the system was corrupt.[9] Sutro's assurance that Comstockers would not have to contribute development money because it would be raised in the East—once thought conducive to the project—when combined with the government franchises, raised suspicions. Alf Doten, in the *Gold Hill News,* called the project a "swindle in the dark."[10]

A number of San Francisco businessmen, who had known him there, also spoke against Sutro and the project. Some mine superintendents began to argue that shafts had passed under the dangerous water belts and no longer needed the drainage. (This was untrue; mines encountered large pockets of water as long as they continued to sink their shafts.) As opposition built, twelve of the twenty-three mines that had subscribed decided two dollars per ton was too expensive and reneged on their commitment. This influenced potential capitalists in the East, and Sutro was unable to raise the three million dollars. His failure to garner eastern support was then used as grounds for the rest of the mines to oppose him: Sutro, his opponents said, was not to be trusted as a financial manager.

The ostensible explanations obfuscated the facts that some companies sought more profitable terms and that there was now a plan to take over the project and squeeze Sutro out entirely.[11] Alpheus Bull, president of the Savage Mining Company, commented that once the Sutro tunnel was defeated, the Bank Ring would construct its own deep tunnel. Sutro replied: "[Bull] thought it was killed off; his indecent haste was so great that he called in the undertaker before the child was dead."[12]

The leader of those seeking to break up the enterprise was Sharon. He and his associates were in the middle of purchasing the mills upon which the Bank of California had lent money and foreclosed. Sutro's master plan involved building new mills on the river near the entrance to the proposed tunnel. The Bank Ring's mills were miles from the site, and either relocating them or transporting ore through the tunnel and shipping it back to their mills was an expensive proposal. Sharon commented publicly that the proposition was "rather a poor prospect for us of this section." There was another reason Sharon opposed the project: he was developing an alternative plan. On May 8, 1867, he filed papers for building a railroad to service the Comstock. The Virginia & Truckee Railroad would haul ore and materials between towns, mills, and mines. Instead of allowing Sutro to become rich, Sharon planned to tap transportation monies for the Bank Ring's coffers. He approached the companies that had pledged to help finance Sutro's tunnel and enticed them to invest in his scheme instead.[13]

There seem to have been several contributory reasons why Mackay opposed Sutro. The primary one was that in 1867 the two mines in which he was primarily involved, the Kentuck and the Bullion, did not have major water problems and did not require a drain. The Bullion, in fact, was an anomaly: it ended up being the only deep shaft on the Comstock that did not encounter large bodies of water. (Centered between rich mines, it also lacked pay ore, and it was surmised that the lack of water was related to its barrenness. It was hypothesized that the rock there was too tight to have allowed the hot water and gasses that carried the rich minerals to the other parts of the Lode.)[14]

The fact that San Franciscans raised questions about Sutro may have contributed to Mackay's decision. He had not yet established a formal partnership with Flood and O'Brien, but he was carrying on some smaller operations with them.[15] Information gathered from their Auction Lunch pipeline certainly would have been passed along to him. Virginia City residents' resentment of Sutro's egocentrism, referred to at times as "insufferable," may also have affected Mackay. The Bullion mine was unique, in that its ownership was made up almost entirely of Comstock residents. In addition, shortly before Mackay affixed his name to the anti-Sutro telegram Sharon bought Walker's share of the Kentuck mine, making him Mackay's partner. Sharon certainly would have attempted to convince Mackay to contest the building of the tunnel.

In any case Mackay joined the other influential mining men to oppose Sutro, delaying the completion of the tunnel for ten years. After it was completed, Mackay said that they "made a great mistake in not supporting Sutro from the beginning," as it was most needed in the 1870s.[16]

The irony in the community action's withdrawing support because of the way Sutro was going about his business is that the tunnel was replaced by Sharon's railroad. The Virginia & Truckee not only milked the populace by perpetrating a funding fraud that cost the citizenry hundreds of thousands of dollars, it completed the Bank Ring's dreaded vertical monopolization of Comstock mining.

The strangest of Mackay's competitive relationships was with his partner, James Fair. They were two of the shrewdest men on the Comstock Lode, and by joining together, they shaped Comstock mining. Fair was among the elite mine superintendents and was for a time a close to Mackay. But Fair was not a man who sustained friendships. It was said that he developed them only until he gained what he wanted. Fair's obituary in the San Francisco *Chronicle* commented that at his death he had a few trusted agents but not "an intimate friend in the world." Mackay, on the other hand, was always thinking of someone else, and at his death eulogies from friends were innumerable.[17]

In his *History of Nevada,* Sam Davis, a contemporary of the principals, wrote of an incident involving Fair in 1886. Fair had served a term as a U.S. senator and was set to run for a second term. Nevada's old warhorse and first senator, William Stewart, was also running. Stewart sent for Joe Goodman, renowned former owner of the *Territorial Enterprise,* to run his editorial campaign. Goodman immediately revived a story about Fair that had been the talk on the street for years. In 1874 Fred Smith, an experienced and popular mine superintendent, had been assaulted and seriously beaten by John Cosser, a miner and prizefighter. Cosser, who gave the spurious explanation that Smith had been talking badly about his aged father and mother, was fined one hundred dollars. Smith never fully recovered from the beating, dying from pneumonia a year later, but before his death Smith filed suit against Cosser and Fair, charging that Fair had hired the professional fighter to accost him.[18]

Goodman wrote up the twelve-year-old story, but the new *Territorial Enterprise* editor, worried about a libel suit, did not run it. He did not need to. Fair heard what Goodman was doing and, rather than contest the allegation, withdrew his name from the Senate race.[19]

Mackay never sought credit; Fair always did. In his book *The Big Bonanza,* Dan DeQuille wrote that Mackay was the "boss" when it came to knowledge of mining. Grant Smith commented: "Dan was a bold man to praise Mackay above Fair when the latter was slyly claiming all the credit for the successes of The Firm."[20]

When Mrs. Leslie wrote of her travels in the West, she spoke of the Big Bonanza and Fair but not Mackay. The Firm's mines had produced nearly twenty-five million dollars in fifteen months, and Fair had shown Mrs. Leslie and her party around the site. Fair described his partners as Messrs. Flood and O'Brien, who formerly "kept a small drinking saloon in San Francisco" and learned of the Lode from a customer in his cups. Fair apparently said nothing of Mackay, giving the impression that while his San Francisco investor partners received equal remuneration with him—each being worth fifteen million dollars—Fair resided on the spot to fastidiously superintend operations by himself.

A year later, in April 1878, a similar report in the *London World* identified Fair as the "manager, superintendent, chief partner and leading stockholder" in the Big Bonanza silver mines. "He has an army of men toiling for him day and night . . . digging, picking, blasting and crushing a thousand tons of rock every twenty-four hours." Calling him the "Silver King of America," the English newspaper said he had dug forty million pounds out of the earth and had another forty million yet to dig. Mackay again went unmentioned.

Mackay's wealth was also overestimated at times, as when, in 1878, England's *Whitehall Review* reported it to be 55,000,000 pounds or $266,000,000—nearly the entire amount of all the Comstock mines. But while Mackay disassociated himself from such media hype, Fair reveled in it, later claiming that he personally had discovered the Big Bonanza, following a thin seam of ore to its source, although it sometimes shrank to a mere "film of clay." This assertion is disproved by contemporary newspaper accounts and is identified by Grant Smith as "a characteristic fairy tale."[21]

An anecdote about the bruising of Fair's considerable pride also exposes others' perception of him. The actress Helena Modjeska visited the Comstock at the height of the bonanza, and the equally famous Fair showed her about underground. Not realizing that her guide with the grizzled beard and canvas overalls was one of the "Bonanza Kings," she laid a fifty-cent piece on a table, thanking him for the tour. Fair stared at her blankly. One of his associates pretended to come to her aid, whispering that he must want more and asking if she could make it a dollar. She laid another half-dollar on the table and walked out, leaving Fair dumbfounded. Afterward in the bars men laughed, betting that Fair took the dollar.[22]

The relationship of Mackay and Fair with the miners who worked for them is revealing. It was reported that the miners liked to see Mackay. If he was dissatisfied, everyone knew it immediately. They dreaded meeting Fair, who was a "sphinx." The men never knew by his actions if he was pleased or dissatisfied. He always treated them as friends, calling them "me boys." But afterward he might ask a supervisor where the man's reduction works were, implying that he was stealing ore; or he might tell the supervisor to fire the man, because he seemed to know too much.

John B. Shaw, who had worked with Mackay in the Mexican mine in 1860, said that he and Mackay used to eat lunch together on the Mexican dump, and Mackay kept him employed for the rest of his life. Regarding Fair, C. C. Goodwin said: "Of course Uncle Jimmie made some millions from [the great bonanza], but it did not change him, rather it made him as the boys on the Comstock said, more so."[23]

Once in the Gould & Curry mine, the elevator cage was not stopped as it rose to the mouth of the mine and workers were injured, being carried into the machinery. Fair attended to the wounded, but when asked what had happened, responded: "We provide every safety that money can buy or anyone can suggest. Those boys don't like to get hurt, but what can we do? There's one thing we can't do with money, and that's to put brains into a man's head." Mackay's attitude regarding the men and the elevators was reflected differently in a story spread about him. When a stockholder thought miners were not worth their four dollars' wages a day, Mackay exclaimed: "Worth it! Why man, it's worth four dollars a day to ride up and down on that wire string!"[24]

Concerning the marital status of miners, when times were hard, Mackay had a rule that married men were given first choice to work. Fair did the opposite: he believed that men with families were less daring than single men and so should be the first released and last rehired.[25] Likewise, Mackay and Fair held disparate views of those injured on the job. Comstocker Will Gillis said: "[Mackay] had a great human heart in his breast, filled with kindness and good will for his fellow men. If a man in his employ became crippled, or otherwise incapacitated for work, by an accident, John would always find an easy job for him in the mill or some place around the works. Mr. Fair was the direct antithesis of Mackay. He had no use for crippled men, saying that he did not want any but whole men in the mine or around the works."[26]

Still, James Anderson, an old friend who had worked for him, thought Fair was "the brains" of the Bonanza Firm and said Fair was "the best friend the working man ever had." And Fair's brother-in-law, who had also been a mine supervisor, said, "Instead of wanting me to reduce the men, he always wanted to know if I couldn't put on more." But observers frequently read ulterior motives into his actions. When Fair was credited with advocating for the four dollar a day wage and miners' rights in 1877, western historian Richard Lingenfelter speculated that he did so only to avoid a work stoppage at a time when the bonanza was reaching peak production.[27]

The Comstock miners represented all western European countries and America, but a list from 1880 shows three nationalities predominated. Of the 2,770 workers, 816 were Irish, 770 Americans, and 640 English. At the beginning of the miners' union movement, Cornish workers had been hired as strike breakers. They had come from the failing tin, lead, and copper mines of Cornwall and were more skilled than those they replaced. These factors, as well as the legacy of bitterness between the Irish and English, caused hostility between the groups. They also lived in ethnic communities: the Cornish in Gold Hill, the Irish in specific neighborhoods in Virginia City. Resentment frequently bubbled just below the surface. Foremen in some mines were clannish, hiring by nationality. A foreman reported that Mackay instructed him: "Treat them all alike. All we want is good men."[28]

Mackay was at ease with employees. He usually knew them by name and greeted them whenever he saw them. Because Fair trusted no one, the mines he superintended had strict rules. All the mines had changing rooms for the miners coming out or entering the drifts. Fair used roll calls to insure no one was unaccounted for at shift changes. He required each man who was leaving to empty the contents of his dinner can into a company box. The men's mining clothes were washed and dried on the premises, and the men had to bathe before changing into their street wear. Fair appeared in the tunnels at various hours, day and night, and anyone found not strictly attending to work or attempting to deceive him in any way was dismissed.[29]

The biggest mines resembled cities underground, with tunnels like streets connecting great open galleries. The heat was oppressive, and the intricate network of timber holding the ceilings and walls in place was tinder dry. Smoking was prohibited, as fire was a constant danger. The ways in which Mackay and Fair dealt with transgressions illustrates differences in their associations with the miners. When Mackay asked an old miner if he had been smoking, the man, certain he was to be fired, responded quickly that there had been a good deal of dynamite blasting that morning. Mackay stared steadily at him and said, "Don't do it again." Wilson Locklin, who worked on the Comstock fifty years, used to smoke on the sly. Mackay overlooked the first offense. "The second time he said, 'Wilson, if you smoke the others will. Now, if I catch you again you'll get your time . . . ' Locklin said: 'That was enough for me. Mackay always meant what he said.'"[30]

One day Fair thought he detected the odor of tobacco in the mine. He continued on his tour but returned to the site a half hour later. Joining the miners, he sat down heavily. "I am surely growing old, a little run through the mine tires me more than a day's work used to. I think if I had a few puffs of a pipe it would refresh me greatly." Various men pulled out pipes and offered them. Fair puffed on one, gave his thanks, and left.

The next day, walking down to the mine, Fair met the same group trudging up the hill. "I thought this was your shift," he said.

"We have been laid off."

"Laid off? That is John [Mackay]. I never get a crew of men that just suit me, that John doesn't discharge them." But the miners knew Fair had issued the instruction and called him names as they climbed the hill.[31]

Through the years the Mackay-Fair relationship deteriorated; there were periods when they refused to talk. In the end they stopped associating with one another. Still, it was the two together who achieved greatness

CHAPTER 5

The Con. Virginia

THE COMSTOCK LODE served the mining industry as a laboratory for developing mining equipment and procedures. Practices utilized there, some introduced by Mackay and Fair, transformed mining into big business requiring large numbers of wage laborers and enormous capital investment. It was the impetus for the first-ever mining stock market, allowing Sharon and others to make far more in stock manipulation than in mining the ore.[1]

Virginia City and the adjoining Gold Hill grew up around the ore deposits to become the first of the great mining boomtowns, but like those to follow they were not intended to last. They were contrary to the small farm communities envisioned by the agrarian republicans who originally promoted western expansion. To gold seekers the land mattered only as long as its mineral resources lasted. Underground workers, along with townspeople providing support services, filled the territory's population roles. Those who ranched and farmed in the neighboring vicinity helped distort the western ideal by expanding their holdings and likewise hiring wage-earning hands to provide for the markets of the towns. Virginia City and Gold Hill were workstations to be utilized until the territory's riches were extracted.[2] (The timberlands in the neighboring mountains were exploited in similar fashion, as their ancient forests were clear-cut.)

In 1860 J. Ross Browne described the "weird and desolate" scene presented by the Comstock to a first-time visitor: "It is as if a wondrous battle raged, in which the combatants were men and earth. Myriads of swarthy, bearded, dust-covered men are piercing into the grim old mountains, ripping them open, thrusting murderous holes through their naked bodies; piling up engines to cut out their vital arteries; stamping and crushing up with infernal machines their disemboweled fragments, and holding fiendish revel amid the chaos of destruction."[3]

On the edge of the arid Great Basin, the territory's water ownership involved equally desperate, convoluted battles. On the Comstock there was not enough river water to power the mills and not enough untainted water for general use. The Carson River needed to be used for fueling the stamp mills, owing to the scarcity of wood. But with the clear-cutting of timber on distant hillsides, winter snows no longer melted gradually to provide a consistent summer flow. When temperatures rose, snow was swept away in grand floods—leaving the river to fail in late summer and fall. It seemed improvident to the milling interests that upstream water was being diverted for ranching and farming.

In 1864 Henry M. Yerington, representing the owners of the Merrimac Mill, brought suit against the ranchers of Carson Valley. The ranchers won by proving they had been irrigating their lands with river water since 1857, before the mills were built. In 1871, a year of largely improved productiveness in the mines but little water, Sharon and his Union Mill and Mining Company, for whom Yerington now worked, filed another suit against the Carson Valley landowners, claiming unreasonable use. The federal court reversed the earlier decision, concluding that the plaintiffs, as well as some of the defendants, had riparian proprietor rights—that is, rights they possessed because they were situated on the river. Those lands located off the river were perpetually restrained from diverting water in canals or ditches, because of the injury it caused to the downstream riparian land users: the mills.

The practical effect of the second lawsuit was that the Bank Ring hired regulators to patrol the ranches and ensure there were no illegal diversions and no wasteful use of water. The watermen posted rules and visited each rancher, insisting that each of them take as little water as possible. If the rancher used more than was deemed necessary, the regulator shut down the irrigation ditches. A regulator named Zeke Edgecomb told about closing ditches: "I mean to say that I took out the dams that raised the water into the ditches. I took them out and filled them up with earth and other matter—filled up the mouths of the ditches, with earth and stones and anything I could get." When a rancher named Banning reopened the ditches after they had been filled, Edgecomb went to the judge in Genoa and got an injunction. Fighting the courts as well as the regulators was too much for the rancher, and his ditches remained filled.[4]

In Virginia City the natural water supply for drinking, washing, and bathing was insufficient and laden with heavy metals. The wells originally used were drained dry as the Comstock population grew. Availability became limited to two questionable sources: Ned "Lame" Foster's distribution of melted snow stored in sacks in abandoned mines carried to buyers on mule back; and seven streams that flowed out of various mining tunnels. Two companies leased the latter, and in 1862 they consolidated as the Virginia and Gold Hill Water Company. Their water, while more plentiful than Foster's, was unreliable, being least available in the summer, when tunnel streams became trickles. It was also mineral laden and unhealthy. Newsman Dan DeQuille noted: "These [waters] were much impregnated with minerals, one of the least feared of which was arsenic. The ladies rather liked arsenic, as it improved their complexion. . . . But there were other minerals held in solution in the water—those that caused diarrhoea [sic] for instance—that were not so well thought of."[5]

Drinking water became an element in the Bank Ring's monopolization of the area. The Ring purchased controlling interest in the $250,000 capital stock of the water company. Citizens who had previously feared they would run out of water now worried about what might happen to costs and about being at the mercy of the Bank Ring.[6]

In 1867 the water company lost control of the stream that provided more than half of the company's total water. The Cole Mining Company procured the rights to the Santa Rita tunnel stream and started their own water company. Prices and availability remained the same, but the citizens were satisfied because another of its enterprises had been wrested from the Ring.[7] The water company began litigation against the Cole Company, but before it was resolved John Mackay and his associates bought out the Bank Ring's interest. Sharon believed he was unloading the afflicted company on amateurs.[8]

The 1871 reorganization of the water company illustrated the unflinching confidence Mackay and his associates had in the Comstock. Their properties produced relatively little that year, and it was a drought year—the driest since pioneers had arrived. In July and August there was not water enough to operate the mills. But the Crown Point and Belcher mines struck bonanzas, and by the end of the year, along with significant developments in the Chollar-Potasi and the Yellow Jacket, Nevada gained predominance over California and all other states in the production of precious metals.[9]

The new water company board was comprised of Mackay, Fair, Flood, O'Brien, and three others: Walter S. Dean, who had been the secretary of several mines; William S. Hobart, who had served in various public offices and was currently the Nevada state controller; and Johnny Skae, a mining speculator who seven years later made a fortune in the Sierra Nevada mine when he duped Fair, among others, into buying during a stock excitement. The new board met in August 1871 and made a radical decision: they would attempt to procure water from the Sierra.[10]

In 1865 a California proposal had been introduced to build an aqueduct out of the mountains west from Lake Tahoe to run the hydraulic mines in the California foothills, irrigate the Central Valley, and provide drinking water for San Francisco. The *Territorial Enterprise* immediately warned those entertaining such thoughts that they would need "twenty regiments of militia" to steal the pure waters of the bistate lake. Although the scheme periodically resurfaced through the years, it was opposed by Nevadans and San Francisco capitalists with Nevada interests, and it never acquired the necessary financial backing.[11]

Mackay and his fellow board members now proposed to bring water out of the mountains near Lake Tahoe east to the Comstock. Transporting it twenty-five miles across the deep depression of Washoe Valley would be an engineering feat never before attempted anywhere in the world. Previously, the greatest pressure under which water had been carried was 910 feet; the new pipe would require withstanding 1,720 feet of pressure. The work would require several reservoirs, including one atop the mountains, another to regulate the discharge into the pipeline, and yet another to collect the water above Virginia City. Hermann Schussler, the engineer who had developed the Spring Valley system that provided San Francisco's water, was sent for to survey the route and determine if it could be done.

Marlette Lake, above Lake Tahoe at eight thousand feet, had already been dammed and used as water storage for a V flume that carried timber down to Carson City. The new proposal included raising the dam at Marlette to stand thirty-seven feet high. Its interior would be earth, its walls masonry rubble. Flumes were needed to collect water from numerous creeks on the side of the mountains and to carry water from Marlette to the second storage reservoir. At one point a four-thousand-foot tunnel, seven feet high, was excavated through granite.

It took almost a year to build the pressure pipe that would carry the water down into and across Washoe Valley. Differing thicknesses of the rolled metal were needed, depending on the perpendicular pressure and lateral curves designed to circumvent hills or rock outcroppings. Each section had to be carefully marked at the factory, as it would fit in only one place in the seven-mile-long completed line. A total of thirty valves were built into the pipes: air valves, to allow accumulations of compressed air to be released at high pressure points and to allow air into the pipes so they would not collapse when drained; and blow-off valves, permitting water or sediment to be drawn off at low points in the pipe. The line would be placed in trenches two and one-half feet deep. Hydraulic engineers thought it impossible to move water so far under so much pressure. The bold scheme was just such a project as Mackay and his associates might undertake. Borrowing great sums of money, they proceeded with construction throughout 1872 and 1873.[12]

Mackay's gambling spirit was displayed in his mine management in 1872 as well. Although he was thought to be nearly a millionaire, his resources were being exhausted. Deposits at the lower levels of the Hale & Norcross became base, and flooding and the necessary replacement of timber framework was increasing the cost of further exploration. The previously overlooked ore in the upper levels had been mined. On July 15 Fair wrote to the mine's secretary in San Francisco, explaining, "The ore which we are extracting . . . is of low grade, and is found scattered very irregularly throughout the vein." He continued, "The upper or old works are not yielding any ore." After a last dividend payment of eighty thousand dollars, assessments were begun. In what must have been anathema to Mackay, Flood made a never-to-be-acted-upon suggestion that miners take a 15 to 20 percent pay cut. Needing a new strike if their success was to be more than transitory, the Firm now invested well over one hundred thousand dollars investigating title and buying rights to the heretofore discredited Consolidated Virginia mine.[13]

The Con. Virginia was created by combining three of six small properties that lay at the north end of the Lode between the bonanzas of the Ophir mine and the nearby Gould & Curry. Ownership of these six properties had been quarreled over since rules for the territory were adopted in spring 1859. Contentions were the result of equal parts ignorance, carelessness, greed, and fraud. One claimant had not posted required notices on boundary stakes. Another had far overestimated his measurement. Others claimed a prior surface location—the sole evidence being their assertion. Locations overlapped, some having been made without reference to the record book. Others were backdated. In early July 1859, a compromise was struck between all parties and possession was assigned.[14]

The surface of the six properties contained one of the largest bodies of quartz on the Comstock Lode. The properties were explored extensively in the early 1860s to a depth of more than 500 feet. The quartz was almost barren of pay ore: the mines, honeycombed with shafts and tunnels, provided only thin streaks of gold at a depth of 250 feet in the White & Murphy shaft and several small bunches of ore in the California mine. The California find was worked out above the 400-foot level and, when the mine was flooded by hot water at 562 feet, all work in it ceased. In 1867 in the *Territorial Enterprise,* Dan DeQuille suggested combining and incorporating the mines, then sinking a shaft "eastward on a line with the new works of the other leading companies." Before the end of the year, the large stockholders in two of the mines bought out a third, and the combination of the California mine, the White & Murphy mine, and the Sides mine formed the Con. Virginia organization, owning 1,010 feet of the Lode.

After months of inactivity, the owners decided to sink a shaft vertically to 1,500 feet, where it might intersect the Lode on its easterly dip. But by the time they began their shaft, in 1869, financing was limited, and by summer 1870 they abandoned the idea at the 500-foot level. Besides financial limitations, deeper exploration was discouraged by the fact that the Ophir mine bonanza had gradually decreased below the 300-foot level, with no significant ore below 420 feet. The Gould & Curry strike had given out somewhat below 500 feet, owing the depth to its greater mass. Many now believed those bonanzas defined the lower boundary of the productive zone in the

Comstock's north end, so the company owning the Con. Virginia explored the 500-foot level, driving a 900-foot crosscut west to the Lode and long drifts north and south. Again, they found only porphyry (feldspar crystals in a dark red, compact mass) and barren quartz. With stockholders having been assessed over $161,000 and everyone discouraged, worked ceased. The Con. Virginia was now commonly referred to as the "forlorn hope of the Comstock." In February 1871 its shares sold for $1.62.

In April 1871 all Comstock mines' values rose in sympathy with the rich Crown Point bonanza. Crown Point stocks quadrupled to $132 a share then continued to rise to a high of $350 in November. The Con. Virginia stocks rose to $18.37 in April but, with no ore to buoy them, fell in May to fluctuate between $6 and $11.50 a share the rest of the year.[15]

The immense Crown Point bonanza was of great interest to Mackay and Fair. It was across the divide from the Con. Virginia, at the south end of the Lode, but it began at the 1,100-foot level. As early as 1865 Ferdinand Bacon Richtofen, a scientist hired by Adolf Sutro, had concluded that the Comstock ore was a fissure vein with a number of parallel fissures filled with deposits along the length of the Lode. He speculated that the ore pockets might extend to much greater depths, concluding that the fifty million dollars previously extracted would be a small portion of the silver ultimately to be taken. William Ashburner, the mineralogist of the California State Geological Survey, concurred. When Richtofen's report appeared, Ashburner wrote: "We have great right to assume that ore exists and will ultimately be found at as great a depth as it is possible to extend underground workings."[16]

But a report commissioned by the U.S. Congress, compiled by two army engineers and a civil engineer who examined the Lode in 1871, concluded that although it seemed to be a true fissure vein, whether the Comstock Lode would continue to be ore bearing was simply a matter of opinion. Mackay and Fair wagered it would, and that some deep repositories were located at the Lode's north end. Late in 1871 Flood quietly began to purchase blocks of Con. Virginia's 11,600 shares. An astute contemporary, Henry DeGroot, later commented: "The leading features of the Comstock vein had by this time come to be pretty well understood. Its true inclination and strike, the shape and pitch of its fertile chimneys, the character of its walls and country rock, as well as the great magnitude of its ore channel and the nature of its contents, had been carefully investigated and definitely settled." DeGroot pointed out that elaborate underground surveys, dissertations, and illustrated diagrams had been published promulgating the doctrine that isolated ore bodies would continue in depth. He commented that Mackay, Fair, Flood, and O'Brien were among those who carefully studied the matter.[17]

At the end of the year, the Firm controlled three-fourths of the Con. Virginia's stock and attorneys had secured title from the various owners. Their work was thorough, including the acquisition of quitclaim deeds from original claimants who had not been associated with the properties in many years. At the January 11, 1872, meeting of the mine's stockholders, Flood and O'Brien were elected trustees, as were Edward Barron (the original backer of Flood and O'Brien); Solomon Heydenfeldt, an original incorporator of the consolidated mine; and broker B. F. Sherwood. Barron was made president. T. F. Smith was named superintendent but was soon replaced by Captain Samuel T. Curtis. The Firm also bought the Kinney, a fifty-foot, unincorporated mine that adjoined the White & Murphy.[18]

It was determined to reach the Con. Virginia's depths by doubling the shaft, beginning from the bottom of the old shaft at the five-hundred-foot level. Bids for sinking the shaft were advertised in the *Territorial Enterprise* on March 31, 1872, but upon further evaluation, it was decided that the machinery currently employed was not heavy enough for work deep below ground, and they withdrew the ad. Instead they approached their old rival Sharon. They wanted to use his Gould & Curry mineshaft at the 1,167-foot level to run a drift north through the Best & Belcher mine and into the Con. Virginia. Sharon agreed, remarking, "I'll help those Irishmen lose some of their Hale & Norcross money." The drift had to be cut eight hundred feet through Best & Belcher mine property before it reached the Con. Virginia. On May 1 work was begun.

At the same time, development of the old workings at five hundred feet was restarted. One reason they had bought the mine was to rework the upper levels. As with the Hale & Norcross mine, they hoped to find low-grade ore to help finance the lower drift. On May 20, at Flood's offices in San Francisco, the Firm increased its number of shares to 23,600, valued on paper at $300 each, raising the worth of its stock to $7,080,000. To insure payment for work in the mine, three assessments, totaling $212,400, were levied during the year.[19]

Early in 1872 the Crown Point bonanza had proven to continue into the Belcher mine (not the Best & Belcher mine mentioned above, but another Sharon had recently purchased). Although its limits were not definitely known, the Crown Point bonanza had proven to be four hundred feet deep, one hundred feet wide, and six hundred to eight hundred feet long. The *Gold Hill News* reported: "It is literally one grand mountain-like mass of rich silver and gold ore, to which all the other bonanzas . . . heretofore found are mere stray nodules."[20]

As with other markets, the prices of Comstock shares were influenced by its leading companies. Speculators were always eager to invest in surrounding mines once a significant strike was made. Further, because the Lode north to south was three miles long, it was thought that a disparate bonanza might be discovered at any time, so in 1872 the entire stock market was driven to fantastic levels. The price of shares of the Con. Virginia—whose company, while owing no debts or liabilities, also had no sign of ore—rose to $60, then $150. Flood hedged his bet, selling large amounts of Con. Virginia shares when it rose over $60.[21]

Flamboyant Sam Curtis, a powerfully built man with a large, bristling moustache, was now superintending the mine. He had been a 49er and a state legislator in California and, having managed several of the Comstock's important mines, was renowned for his luck. Above ground his life was adventuresome, but he spent much of his time underground. Although given to extravagance when managing a mine, he had knowledge of the stratification, crystallization, and other characteristics of the Lode similar to that of Mackay and Fair. His experience and work habits go a long way in explaining his luck.[22]

In spring and summer 1872, Curtis's exploration at the Con. Virginia's five-hundred-foot level was in turns promising and discouraging. When the Crown Point boom dissipated, the price of Con. Virginia stock again fell, settling at below $30 a share. Throughout the summer in its "Local Mining Summary," the *Gold Hill News* spoke of promising crosscuts or average improvement or quartz of "a favorable character." But in one instance regarding the quality of the quartz, it reported, "we cannot speak with certainty"; in another it was called low-grade ore; and sometimes the Con. Virginia was simply left out of the report.[23]

Dan DeQuille commented that miners see no farther into the ground than anyone else, comparing a man at the bottom of a shaft to one groping about in a dark cellar. "He knows which way to go to reach the vein, but when once he is in the vein he may almost touch that of which he is in search without finding it." When a friend later asked Mackay whether there were laws that reveal veins, he replied: "There is no law in mining but the point of the pick."[24]

At the 1,167-foot level, the Con. Virginia drift was dug northerly. Although said to be prosecuted as quickly as possible, it was slow going: the miners were working with hand drills, swinging three-pound hammers in one hand and twisting the drill between strokes with the other, gaining on average one foot an hour; or two-man teams utilizing the dangerous technique of "double jacking"—requiring one man to wield a seven- to ten-pound sledgehammer, while in the dim candlelight another twisted the drill and held it for the next strike. Double jacking averaged one and a half feet per hour.[25]

Early in the exploration there was no water in the drift, having been drained by the Ophir and Gould & Curry shafts to the north and south. Numerous quartz stringers and small ore bunches indicated larger deposits to the west, and a number of crosscuts were begun toward the Comstock Lode. But the ground to the west had not been drained. It was saturated with hot water. Each of the crosscuts had to be abandoned shortly after it was begun, and the main drift was bulk headed to prevent the water and muck from flooding it. As was their wont, Fair and Mackay built the mine securely as they went. The deep drift was firmly timbered, and solid planking was laid throughout.[26]

At the end of July 1872, the *Gold Hill News* said indications were "exceedingly flattering." On August 3, although it was reported that there were no rich ore bodies, prospects at both levels were said to be excellent. In mid-August Curtis came to a heavy clay wall, leading mining reporter Doten to comment, "An important development is expected." These announcements were not uncommon in various mines, and no one outside the Firm seemed to pay attention; stock prices remained below $30. But Mackay and Fair ordered new hoisting machinery, including a state-of-the-art one-hundred-horse-power hoisting engine. When the northerly drift was 178 feet into the Con. Virginia, Curtis turned it to the northwest and, on September 12, 1872, he came to a seven-foot-wide fissure of quartz, porphyry, and clay that held low-grade ore. Following the cross fissure, Curtis continued westward toward the Lode through the unstable ground.[27]

The find spurred further action by the Firm. Flood immediately bought back the shares he had sold, causing stock prices to rise from $29 to $57 between September 13 and 19. Doten, at the *News,* sensed something was up. Although the Firm was generally acknowledged as having been much more open than other owners about keeping the public informed of developments (those in control usually kept improvements secret so they might secure quantities of stock for themselves), in this instance they played things close to the vest. Doten commented: "The proprietors and managers of this mine keep their own counsel and are not inclined to be at all communicative. They must have a good thing."[28]

At this time the company also moved to buy the last two of the six original mines, the Central and the Central No. 2. Because most ore bodies on the Comstock pitched southward, they also bought the adjoining Best & Belcher, which never produced an ore body, and Sharon's long-unproductive Gould & Curry. On September 19 they began grading to build new, first-class hoisting works. The bed of masonry for the new engine was 22 feet deep. The main building was to be 104 by 45 feet, with a 40 by 40-foot boiler room at the west end and a carpentry shop and a blacksmith shop at the east. A large number of men were employed in the work.[29]

Curtis now publicly proclaimed his confidence in finding good ore in the ground, saying that when properly opened, it would prove to be one of the best mines on the Comstock Lode. It was decided to sink the shaft from the Con. Virginia 500-foot level to 1,200 feet to meet the drift. Work at the 500-foot level was abandoned in order to concentrate on the deeper tunnel. Curtis continued following the fissure, which now led him northeast. Around this time Fair left, going to winter with his family in Oregon, and Mackay went on a three-month trip to Europe to spend Christmas with his family.[30]

Mackay's extended family was enjoying life in a fashionable Paris hotel. As Eva recuperated from an operation on her hip, they were immersing themselves in French culture, attending the opera, the theater, and the Louvre. Mackay told Louise that nearly everything taken out of the Hale & Norcross was being sunk into the Con. Virginia. Sharon, he said, was calling them the joke of the Comstock. She asked him what they would do when all the profits were gone. The answer must have terrified her. He said they would go broke and start over.[31]

When Mackay returned to the Comstock, work at the mine was progressing. By February 1, 1873, the new hoisting machinery was in place and tested, running quietly without jolts or jarring. The drift had cut through several stringers of quartz that showed promise, but the work was pushed through without attempting to remove them.[32]

On the last day of February, the Con. Virginia shaft was 710 feet deep. It was being worked day and night and sank 3 feet a day. Completion would allow fresh air to blow through to the long drift at the 1,167-foot level. Large blowers had been installed in the drift, carrying air from the Gould & Curry shaft, but they were inefficient, and the air at the drift face was hot and growing unbearably foul. The tunnel, following the fissure, turned and twisted, which also hindered the circulation of air. The fissure, surprisingly, now ran northeasterly—away from the Comstock Lode.[33]

On March 1 a substantial body of ore was reported. Mackay and Curtis began removing twenty-five tons daily through the Gould & Curry. Three good deposits of ore were now passed through, saved for later extraction. By March 20 the tunnel ran 50 feet further, and some of the ore in it was very high grade. Assays of the best ran between $300 and $400. Others averaged $179 a ton, leading the *Territorial Enterprise* to state that the mine was on the verge of developing a large, very rich and permanent deposit. On March 22 stocks closed at $61.[34]

On March 23 a disagreement between Mackay and Curtis led to the removal of Curtis as superintendent. Dan DeQuille protested the popular man's firing in the *Enterprise:* "John Mackay took charge of the mine last Sunday morning, superseding Captain Sam Curtis. We have heard of no reason for the change. Captain Curtis seems to have been doing all that could have been done by any man for the development of the mine."[35] Characteristically, Mackay did not make public his reason for removing Curtis. He supervised work in the Con. Virginia himself until Fair returned to be reelected superintendent.

At the end of March, with no further report of a strike, stocks again ebbed, to close at $53½. The mine was drifting underneath buildings of the town, and the fact that its stock prices would languish for the next six months adds poignancy to Dan DeQuille's later comment: "For fourteen years men daily and hourly walked over the ground under which lay the greatest mass of wealth that the world has ever seen in the shape of silver ore, yet nobody suspected its presence." Of this specific period, he said: "Although people knew in a general way that there was an abundance of rich ore in the mine, they did not get excited about it, nor did they trouble themselves much about it in any way, further, perhaps, than to say: 'Well, I am glad to hear that the Consolidated folks have a big body of ore; it will be a good thing for the town.'"[36]

The sinking shaft was now at 784 feet, and there was little patience for its completion. In the main drift crosscuts had been practically suspended, since as soon as workmen got a few feet from the main drift, the heat became intolerable. By mid-April the Firm had four shifts of men working six hours each in the sinking shaft, so as to keep them fresh and hasten the work. Generally in vertical shafts, three or four compartments, perhaps five feet by six feet, were developed: one sinking compartment for excavating, one or two for hoisting and extracting ore, and one for pumping out water. For the time being the pumping compartment was abandoned so work on the others could be expedited.[37]

Early in summer Mackay returned to Paris. Louise, at twenty-nine years old, had taken to the life of an American gentlewoman. John Jr. was nearly three years old. Stepdaughter Eva's limp was less noticeable, and Mackay was said to be extremely fond of her. Mackay told Louise what he had only hinted at in letters: the Con. Virginia was rich in ore. It was now simply a matter of time. He felt certain that their risks in preparing the mine as thoroughly as they had was about to pay off. The couple made a plan: although Louise's parents and young sister would remain in Europe, after John returned stateside she and John Jr. would follow and take up residency in San Francisco. He would then split time between the Comstock and the city.[38]

During the previous two years, all the elements of the Virginia and Gold Hill Water Company's great pipeline from the Sierra to Virginia City had been completed. It took 6½ weeks to lay the pipe across the valley and its thirteen deep gulches and up the west slope of the hills outside the town. The board members had borrowed heavily to complete the work; the cost was now over two million dollars. When sections of the line were tested, lookouts signaled success or a break in the line by use of smoke during the day and fire by night, learned from the Paiute and Washoe Indians. Only one sheet of metal and three sleeves (cast-iron bands over each connecting joint) proved faulty. On the night of August 1, 1873, water was released, and its progress was traced as the air valves were forced open to allow the compressed air to escape before it. DeQuille commented that compared to these shrieks, "the blowing of a whale was a mere whisper." When the pure mountain water gushed into Virginia City, cannons were fired, bands paraded, and rockets were launched all over the town. Henry R. Whitehill, Nevada state mineralogist, said: "One of the most needed and beneficial undertakings for the prosperity and health of Virginia has been accomplished in the completion of this company's water flume from the Sierra Nevada, whereby the city and mines are supplied with pure water from the mountains." If they had not been previously, Mackay and his associates were now heroes throughout the Comstock.[39]

In August, with the Con. Virginia shaft at 1,100 feet, work was stopped so the considerable infrastructure work could be completed. A massive escarpment wall for support of the ore bins was built, and the tramway from the mouth of the shaft was completed. Pits for the pump tanks were excavated at the 250- and 500-foot levels, oversized tanks were built, and the great pumps were lowered.[40]

With Mackay back on the scene, the Con. Virginia's main vertical shaft was finally connected at the end of September, and on October 1, 1873, ore began being hoisted. Shortly thereafter, two hundred tons were being milled daily in the Mariposa and Bacon Mills, and a third, the Occidental Mill, was being brought on line. The drift, having advanced to a spot 250 feet southeasterly, now struck an exceedingly rich deposit of ore.[41]

The price of stock, already $100 a share, now surged to $240. Usually, when asked about stock investments even by friends, Mackay demurred, saying he knew nothing about the market. There are only two recorded instances wherein he encouraged friends to invest. Both concerned Con. Virginia stock, and both ended unsatisfactorily for the advisee. When he urged his friend Dolph Hirschman to buy stock, Hirschman refused. He and his two brothers had been investors in the Bullion mine, which Mackay had superintended earlier. After six years of assessments and heavy losses, Hirschman had taken an oath to never invest in stocks again. Refusing to buy Con. Virginia stocks set him up to be second-guessed the rest of his life.

Mackay was also fond of Hermann Schussler, who had engineered the water pipeline. Mackay suggested early that he buy Con. Virginia stock. He bought $12,500 worth, but a friend advised him to sell, believing that the Firm was attempting to get back the fees they had paid him. Schussler related what happened after he sold his shares for $50,000 and took his family to Europe: "I had been in Paris only thirty days, when I received a cable, 'You are offered one million five hundred thousand dollars for your Cons. Virginia. Will you take it?'"[42]

On October 18, 1873, the Con. Virginia trustees met and, proceeding in typical Comstock-mine fashion, increased the shares of stock. The number was raised from 23,600 to 108,000, with the Firm owning 62,000 shares.

Flood said it was done to keep stocks at a moderate level and "give the little fellows a chance." In an article 14 months later, the editor of the *Mining and Scientific Press* commented that the increase in stocks was stock jobbing, pure and simple: " 'Giving poor men a chance' is too thin an excuse to be swallowed by anybody." Regardless of motive (and Flood is generally credited with being "more open" than other mine owners), the increase in shares improved Con. Virginia capital stock from $7,080,000 to $10,800,000 and dropped the price of a share to $48.[43]

On October 29, 1873, in the *Territorial Enterprise,* Dan DeQuille reported on exploring the "long forbidden lower levels" of the Con. Virginia. Because of his suggestion years earlier about mining the property in depth, and because of his unquestioned status as the territory's leading reporter on mining, Mackay and Fair gave him the exclusive right to be the first outsider to evaluate the mine. Rumors had been rampant, with thousands of guesses both favorable and unfavorable, and the owners resented some of the negative publicity they were garnering. What DeQuille found was a chamber at the 1,167-foot level, 20 feet high and 54 feet wide, with masses of ore on all sides. DeQuille was urged to take specimens from various parts of the deposit. The drift had traversed a 200-foot-long vein of ore and crosscuts had opened it, so its extent was fully revealed. DeQuille's samples were "wonderfully rich," assaying $632.63 at highest and $93.67 at lowest. But the development was tempered by his comment, "But of course we can see into the ore deposit no further than the openings have been made." Because the strike was in an unexpected place, and perhaps influenced by DeQuille's uncertainty regarding how much farther it might reach, people reacted in restrained fashion. Stocks sold for $51 on the 29th and for $51.25 on the 31st.[44]

The information not released was that—as in all good mining operations—several prospecting winzes had been sunk to find out whether the ore was "going down," and it was.[45]

CHAPTER 6

Bonanza Silver
and Kings

O N   A U G U S T 1, 1873, the same day the first pure water ran through pipes
from the Sierra to Virginia City, Andrew Smith Hallidie's cable car was tested
on runs up and down a hill on Clay Street in San Francisco. His invention
gave the city an attraction and signature symbol recognized into the twenty-
first century. A short time later, not many blocks away on O'Farrell Street,
Mackay paid the regal price of thirty-one thousand dollars for a three-story
wooden house across from a row of narrow, two-story homes. A pregnant
Louise Mackay and her two children had returned to San Francisco from
their extended stay in Europe. A city developing a modern mode of transportation was the sort of place that would appeal to the forward-thinking
young mother. She did not want to return to live on the Comstock, saying
she knew the place too well.[1]

Louise Mackay had grown up in a mining camp and supported herself as a seamstress in Virginia City. She wanted something better for her children and herself than living in the male-dominated mining society. The Comstock environment was hostile—snow in winter, winds and dust in summer, denizens openly caroused, and violence was ever present. Clarence H. Mackay was born in San Francisco, April 17, 1874. In Virginia City on April 18, a shoemaker shot his wife in the face during a domestic dispute then killed himself. Two weeks later, two miners were killed and four injured when an ore cart broke loose from its cable, and there was a second case of a man shooting a woman, then himself—this time jealousy was the motive. Additionally, divorce was steadily increasing on the Comstock, and all classes of women were using opiates, some prescribed by physicians, some not.[2]

San Francisco had its strange and exotic quarters, including the Barbary Coast—three solid blocks of dance halls and depredation, where some Kanaka Pete could be found shooting up the Eye Wink saloon because of Iodoform Kate. But the city was also cosmopolitan, with fine new mansions, gardens, chic dinners, and courtly balls. The nouveau rich had developed an elegant society, described by Amelia Ransome Neville as "something of the legendary hospitality of Spanish days blended, with that of the old South represented by so many families, into a manner of living indescribably generous and delightful, distinctively Californian."[3]

While she lived there, Mrs. Mackay brought a grandiloquent lifestyle to her neighborhood. One neighbor commented on the huge cans of milk delivered for Louise's bath. Another told how a hired carriage often stood in the street at the bottom of the broad, high staircase. "Wooden coachmen sat in aristocratic immobility, footmen sprang like acrobats from their seats to hand out the ladies of the house." When Louise hosted a ball, crimson velvet carpet ran from the curb to the entry of the house. A friend met Mrs. Mackay, baby Clarence in her arms, and her mother, Mrs. Hungerford, in the Grand Hotel. She said that Mrs. Hungerford was most enthusiastic about the family's good fortune. The same friend commented that Louise did a good deal to help the Mackays' friend Dolph Hirschman with his jewelry business, which he relocated from Virginia City to San Francisco. Another friend reported that when shopping, "Mrs. Mackay never asked the price of anything." While the Mackays were not yet among the Burlingame Country Club elite, Louise felt her time was coming.[4]

The foremost figure in San Francisco society, as well as its business leader, was William Ralston. He invited the Mackays to at least one soiree at his extravagant and courtly Belmont estate.[5]

Ralston was William Sharon's partner and was the power behind the Bank of California. The bank's sphere of influence stretched from the Pacific to the Rockies. Its board of directors was composed of the heads of San Francisco's largest mercantile, manufacturing, and transportation firms. From its inception, the bank was the commanding financial and political entity in the Far West.[6]

Ralston used bank money to finance many of the projects that were building California. His nod of approval allowed entrepreneurs, scientists, and dreamers to secure money for their projects. Real estate; the West's first vineyard; a sugar refinery; a watch factory; a furniture company that utilized only California wood; a manufactory for building railroad rolling stock, cars, carriages, and agricultural equipment and another for making locks; irrigation works in the Central Valley; and a woolen mills company were among the businesses subsidized by Ralston and the bank. In January 1872 the *Commercial Herald & Market Review* wrote about a wheat deal with which he was involved: "The name of Mr. Ralston gives to the enterprise the strongest possible assurance of success. Wherever the Bank of California is known—whether in Europe, Asia, or America, this gentleman is recognized as a leading power on the Pacific Coast." Ralston himself had conceived the Bank of California as well as San Francisco's avant-garde New Montgomery Street, the Grand Hotel, and the California Theater, and he was engaged in building what would be the unrivaled Palace Hotel. His country estate at Belmont was a princely estate where he hosted presidents, monarchs, emperors, and socialites.[7] But Ralston was equally at ease with the workers in his factories and the construction crews at his projects. And he was a friend to Mackay, Flood, and their partners, who had large accounts in his bank.

On the Comstock in December 1873, the Con. Virginia shaft reached 1,300 feet. Drifts dug south and east were rich with ore. The Firm now organized a second company, the California Mining Company, conveying the old California claim to combine with the northerly properties: the Central mine, Central No. 2, and the Kinney claim, encompassing a total of 600 feet. That property, called the California mine, would be developed concurrent with the ongoing operation. At the end of the year, the Con. Virginia had yielded $645,587. Its shares were being sold for $80, the equivalent of $400 at the previous number of shares.[8]

By February the shaft was at the 1,400-foot level, where the main body of ore was similar in shape to the ore 100 feet above. There was also a large L-shaped body 50 feet by 150. And the vein at this level was richer, at $54 a ton. With the ore increasing in volume and richness as the shaft sank deeper, the only question was how many times over a millionaire each of the four Firm members would be.

On February 6, 1874, a transaction that would have dramatic ramifications took place. Joseph Goodman sold the *Territorial Enterprise*, including the building, equipment, and property, for $50,000, five times its worth, to William Sharon. Five trustees were named to the newly incorporated Enterprise Publishing Company, including Mackay. The five were identified in the press as friends of Sharon, and rumor had it that the change was made in order to further Sharon's U.S. Senate aspirations.[9] Sharon had run for the Senate two years earlier and had been defeated in large part because of Goodman's energetic opposition, which included daily attacks in the *Enterprise*. Sharon promoted the powerful and cagey associate editor, Rollin Daggett, to the editor's post and, with Daggett carrying his banner, began what would be a successful Senate campaign. Within a year Sharon's friendship with Mackay would be strained to the breaking point.

Men who struck rich bodies of ore on the Comstock dealt with success in a variety of ways. The picturesque early locators hurriedly sold their claims to unsuccessfully explore in other remote and desolate areas. Some early bonanzas were wasted on luxurious, monumental mills and outbuildings, displays of conspicuous wealth. The Gould & Curry president complained: "Every stockholder wanted [the ore] snaked out at once, at any cost, and so we wasted a third of our profits." Men like George Hearst, the future U.S. Senator from California; Lloyd Tevis, future president of Wells Fargo & Company; and Thomas Bell, who became a baron of quicksilver production, took up lives in the more comfortable environs of San Francisco, investing their profits in other industries. Some insiders sold out at the first unfavorable developments. Mackay's partner J. M. Walker sold as soon as developments in the Kentuck mine became favorable. J. P. Jones, later joined by Sharon and Fair, claimed his prize as member of that most exclusive club, the U.S. Senate. With the richness of the Con. Virginia strike now assured, Mackay and Fair took a different tack, continuing to build the infrastructure of their mines.[10]

A drift was dug from the 1,500-foot level of the Gould & Curry to connect with the Con. Virginia's north winze, reducing the temperature at that depth and circulating clean air. The rock at 1,500 feet was exceptionally hard, and so, for the first time in one of their mines, Mackay and Fair introduced the use of Burleigh drills. The drills were cumbersome and heavy, requiring two men to operate them, even though they sat on a trip rod or on a vertical bar between two jackscrews. Compressed air powered a piston to which a drill bit was attached. Utilizing fifty-five to sixty pounds of pressure per square inch, the drill attained two hundred strokes per minute.[11]

The Firm began sinking a shaft of three compartments 1,040 feet east of the Con. Virginia shaft. It would be used by both the Con. Virginia and California mines and, utilizing the first initials of each, was called the c & c. The ore body was being explored with drifts, crosscuts, winzes, and "up-raises." It continued to spread out and become richer as it gained depth.

On May 11 the Con. Virginia declared its first dividend of three dollars a share, and a month later the Firm began building an immense, state-of-the-art stamp mill just below the mine. Once put into operation seven months later, it used 24 cords of wood daily to reduce 260 tons of ore. Each stamp weighed eight hundred pounds, and they were deafening when in operation.[12]

Con. Virginia stocks fluctuated, in an ever-increasing volume of business, throughout 1874. Although the general nature of the strike was known, its dimensions and richness were matters of conjecture, and so the market rose and fell unevenly with each fresh development and rumor associated with it.[13]

In October, explorations began on the 1,500-foot level. The ore body had increased in size, and its richness was unmatched. The general public was now allowed underground to judge the mine for themselves, but the reports issued by visitors spread excitement throughout Virginia City and San Francisco. The *San Francisco Chronicle* commented that while a history of heartless frauds by greedy mine owners would fill a book, "we hear of scores of poor men and women who, getting information directly from the managers and leading owners [of the Con. Virginia], have made fortunes by the advance in price." Stock prices, previously hovering near $85, rose to $110 in mid-October.[14]

In European, Mexican, and South American silver mines, veins were typically a few feet in height and width. The 1,500-foot level of the Con. Virginia was the base of what was now projected to be a 350-foot mountain of silver streaked with gold. Mackay later used a childhood memory to describe what was being called the "Big Bonanza": "[It was] about as high as the steeple of Trinity, and in area as large as the City Hall Park of New York." Dan DeQuille reported that assays from places of concentrated ore reached $5,000 to $10,000 a ton.[15]

In the southern-most end, a large stope was being dug from the bottom up, supported by square-set timbers as the work progressed. It was a solid mass of silver ore extending over 150 feet in width. DeQuille reported: "On all sides of the pyramidal scaffold of timbers to its apex, where the candles twinkle like stars in the heavens, we see the miners cutting their way into the precious ore." At the northern end, fourteen feet south of the California mine line, a chamber ten feet square was dug, "the walls of which were a solid mass of black sulphuret ore flecked with native silver, while the roof was filled with stephanite, or silver in the forms of crystals . . . and the masses of ore taken out were almost pure silver. . . . Look where you might, you saw but a solid mass of black sulphuret ore mingled with the pale-green ore containing chloride of silver." The generally matter-of-fact state mineralogist's report called the vein "marvelous" and "unprecedented" and said that the refining capacity of the Pacific Coast was scarcely sufficient to handle it. Ore from the chamber had an average assay of $600 a ton. By the end of the year Con. Virginia stock rocketed to $580 a share.[16]

Virginia City was now exhilarated, famed throughout the world, and near the height of its prosperity. On the streets no one knew who might be the next to strike it rich. Stockbroker George Mayre Jr. commented, "Life there for a time had a singular interest and charm, there was a suppressed, semiconscious excitement and expectancy which was not confined to any set of persons but was shared by the entire community."[17]

San Francisco reveled in the wealth of the citizens of the West. On January 4, 1875, the *San Francisco Chronicle* listed a dozen multimillionaires, including Sharon, Ralston, J. P. Jones, E. J. "Lucky" Baldwin, and Leland Stanford, commenting: "Worth $5,000,000? Well, yes, they may be worth that paltry sum. . . . These are only our well-to-do citizens, men of comfortable incomes—our middle class. Our rich men . . . are Mackay, Flood, O'Brien, and Fair. Twenty or thirty millions each is but a moderate estimate of their wealth."[18]

In becoming the richest of those who made fortunes on the Comstock, Mackay earns a place as the third individual who, during three distinct periods, dominated the Comstock scene. Nevada's first full-term U.S. senator, attorney William Stewart, towered over the territory during its formative stage; banker and developer William Sharon organized the mining industry and built a vertical monopoly to control it during its middle period; and the "honest miner," John Mackay, now became the predominant figure in the last of its prime years. The three differed in many respects, with Stewart garnering personal and political power; Sharon building fortunes for himself and his cronies in the Bank Ring; and Mackay, with the help of Fair and Flood, wresting power from the Ring, allowing a more evenhanded power structure.

The role that taxation of the mines played in the development of the territory, as well as in the rise of the three power brokers, is of interest. It is even more significant when one recognizes the role it played in initiating one of the great tragedies of the era, the failure of the Bank of California and resultant death of William Ralston.

William Stewart was a large, powerful man who made his mark during a decade of serving as a mining-camp lawyer in California. He proved himself against mob injustice in the camps and bullies—including the infamous killer Long-haired Sam Brown in Virginia City—and he bribed juries, consistently winning substantial fees for his efforts.[19] When the question of Nevada statehood was considered in 1863, Stewart pushed hard to support the cause. He was seen, though, as being the agent of San Francisco capitalists who were vying to control the territory's mines. Largely because of opposition to Stewart and those he represented, the first attempt at passing a state constitution failed.[20]

But during that first convention, Stewart argued that taxes should not be levied on unproductive mines, but only on the net proceeds of those that were producing. This position eventually won the day, when businessmen realized their prosperity was tied to the investments of the wealthy mining interests. The constitution was approved, and Stewart was elected to the U.S. Senate. His initiative exempting mines from taxation until they became profitable became the law of the land.

In March 1865 property taxes for each $100 were set at $1.50 for the county and $1.25 for the state, but the tax on mine proceeds was limited to 50 cents for the county and 50 cents for the state. The law also stated that $20 of each $100 of mining profits could be deducted for working the ore and that only ¾ of the remaining amount could be taxed.

As William Sharon and the Bank Ring gained power and political influence, provisions were introduced to further reduce the liability of mine
owners. In 1867 a special session of the state legislature, reportedly called at
Sharon's behest, cut the tax in Storey County, where his Comstock mines
were located, to 25¢ per every $100 of profit. (Taxes in other counties and the
state tax remained at 50¢.) A further break was given when it was determined
that $40 a ton could be exempted from ore that had to be worked in the
smelting process and $18 a ton on free ore. This caused Myron Angel, editor
of the first Nevada history in 1881, to comment: "In this respect the new law
out-Heroded Herod himself."[21]

In 1871, with Sharon now in almost complete control of Nevada politics
and a bonanza having been struck in his Belcher mine, a provision passed
the legislature that allowed a deduction for mine proceeds equal to the cost
of extracting the ore and converting it into bullion. Deductions for production were allowed at 50 percent on ore valued at over $100 a ton, 60 percent
at between $30 and $100, 80 percent at between $12 and $30, and 90 percent
on ore valued at less than $12 per ton. Under the new law, owners smelting
ore worth $40 a ton might figure their expenses so that only $1 a ton was li-
able for assessment.[22] Perception on the street was that Sharon and the Bank
Ring were making out like bandits. The perception was accurate.

In January 1875 Sharon was elected to replace William Stewart in the U.S.
Senate. The Virginia & Truckee Railroad, constructed at taxpayer expense,
was paying him an income reported to be not less than $2,000 a day.[23] In the
past he had wanted mining taxes reduced so his own funds were not used
to pay off the railway bonds. The indignation of ranchers, businesspeople,
and other taxpayers of Storey County had reached a high pitch over the
tax discrimination. Sharon's Belcher bonanza was showing signs of playing
out, so he no longer insisted on the great inequality in the tax code. When
a proposal was introduced in the legislature to assess Storey County's mine
proceeds at $1.50 per $100, the same as every other property, Sharon did not
oppose it. On February 20, 1875, the bill passed both state houses in nearly
unanimous votes.[24]

In 1866 taxes for all Storey County mines totaled $17,772. With the tax
law change in 1875, the tax levied on the Bonanza Firm alone would average
$250,000 per year over the next five years. It was but a fraction of the Firm's
yearly average gross yield of $17,700,000, but Mackay and his associates be-
lieved it unfair and blamed Sharon. It was galling to them that, having bested

him in the mines, they were now required to pay an inordinate proportion of the debt on his cash cow, the railroad. They declared war against Sharon.[25]

Mackay did not possess the wiles necessary to be involved in politics. The Bonanza Firm's attempts at confronting the new tax legislation were clumsy and ham-handed. Believing the law unconstitutional, they refused to pay any tax to county or state. The fight over payment that the Bonanza Firm would lose, resulting in their having to pay the full amount of the taxes, would be engaged in the legislature and the state courts. Over the next four years, the governor and the U.S. Supreme Court would also become involved.[26]

Since gaining control of the Hale & Norcross mine, the Bonanza Firm had utilized their own stamp mills. On August 14, 1874, they had incorporated the milling operation as the Pacific Mill and Mining Company. By 1875 they were using nine mills, including the massive Con. Virginia sixty-stamp mill. In 1875 they also began manufacturing their own lumber and supplying their own cordwood for fuel. They soon employed two hundred men in two large saw mills, capable of manufacturing forty thousand feet of lumber per day.

John B. Hereford, the Firm's superintendent of mills, had spent two years quietly buying up twelve thousand acres of timber eleven miles above Reno. A fifteen-mile flume was being built at a cost of $250,000. Some of the trestle work was higher than the trees and built in a fashion substantial enough to support a narrow-gauge railroad. The timber operations were meant to supply the Firm's mines with timber and their stamp mills with the forty thousand cords of wood needed yearly. All this gouged Sharon's Bank Ring operations, which had previously monopolized the Comstock's wood industry.[27]

A news blurb in April 1875 caught the tenor of one skirmish between the adversaries, saying, "There is considerable strife between the Mackey-Fair [sic] crowd and the Bank of California for possession of the water rights in Dog Valley, near Verdi." Another item announced a more serious menace to Sharon's fortune: Mackay, upset with rates Sharon was charging, was rumored to be considering building his own railroad. The editor wrote that the project might be completed within the year. Sam Davis reported that the railroad people reduced rates at once, "for they knew that when Mackay said a thing he meant every word of it."[28]

The Mackay railroad was never built, but another element of the industry to be contested would have ramifications beyond anything expected. On

May 28, 1875, the *Nevada State Journal* announced: "Articles of incorporation of the Nevada Bank of San Francisco were filed on Monday in San
Francisco. Object: 'To engage in and carry on the business of banking to
such an extent and in all such branches as may legally be done under the
Constitution and laws of the State of California.'"

Named as bank directors were Flood, O'Brien, Mackay, Fair, and former
naval officer Louis McLane, who became bank president. McLane had been
a major in the Mexican-American War, one of the original incorporators of
the Bank of California, the president of the Pacific Mail Steamship Company,
and the president of Wells, Fargo & Company. He was able in management,
followed rules, and was said to be severe in discipline. Capital stock for the
venture was five million dollars in gold, the first bank in the world to open
with so much bullion.[29]

The Bonanza Firm having successfully challenged the Bank Ring cartel
and won control of the best mines and the subsidiary businesses (except
transportation—over which Mackay exerted considerable influence), the
Firm now set its sights on assailing an invincible institution. They meant to
do no less than to engage the Bank of California in a fight to the finish.

CHAPTER 7

1875

In January 1873 the Crédit Mobilier scandal broke over the U.S. Congress. During construction of the Union Pacific Railroad line, the Crédit Moblier construction company received the proceeds from government bonds, securities, and land and town-site sales. There was no accounting for forty-four million dollars of those monies. House member Oakes Ames had distributed shares of the railroad's stocks to the vice president, the vice president–elect, cabinet members, senators, and congressmen, so that when issuing advise or voting, they would protect their own property. Discredited and soon to be impeached, Ames stood in the House chamber and read the names of congressmen and senators whose votes he had bought.

The exposé was the latest disgrace perpetrated during the Gilded Age. It followed Boss Tweed's one-hundred-million-dollar robbery of New York City, Cornelius Vanderbilt's railroad plundering, Jim Fisk and Jay Gould's railroad stock watering, Collis Huntington's bribing of California legislators, and John D. Rockefeller's chicanery and strong-arm tactics in building Standard Oil. In the Crédit Mobilier outrage, excepting Ames, the congressmen were not censured, because they claimed they had not known Ames's intentions. Thousands had lost fortunes when the Union Pacific was looted, and distress spread through the Midwest. When the expansion of agriculture in the West and a concurrent drop in world prices started a stock-market panic, the depression would last for years.[1]

On the West Coast the repercussions of the depression were not felt until 1875. Industry, the railroads, and shipping were sound; real estate fluctuated but remained a steady market; and agriculture production was strong. Of course all the economy west of the Rockies was dependent on mining and the mining market. When stocks rose, so did bank deposits; when they fell, money was drained. Now, the value of silver was undergoing a significant change.

In 1873, after two years of debate, Congress passed an act dealing with assay offices, the mints, and coinage. The act left the silver dollar off its list of coins, and so, following the lead of Germany and the Latin Union, moved the country from the double standard of gold and silver to the gold standard. As more and more silver was discovered, the demonetized metal's value shrank. Early in 1873 it sold at $1.32 an ounce. By August 1875 it fell to $1.21 an ounce. (It continued its slow decline, and by 1886 sank to $1—despite passage of the Bland-Allison Bill in 1878, which required the government to buy between two and four million ounces of silver a month.) After 1873 California banks were weakened, because of the decline in value in that part of their reserves held in silver.[2]

In October 1874 the Con. Virginia strike, now established as being worth tens of millions of dollars, had caused wild excitement on the San Francisco Stock Exchange. A body of ore had been discovered in the Ophir mine, adjacent to the California. It extended from the 1,435-foot level to 1,700 feet and was assumed to be part of the Con. Virginia bonanza.

As was often the case, William Sharon knew to invest in Ophir stocks before anyone else. Apparently, mine superintendent Sam Curtis had surreptitiously passed the news to Sharon. The banker bought something over 20,000 of the mine's 108,000 shares in August at $20 a share, before anyone else knew of the strike.

In September the stock sold at $52, and Sharon engaged James Keene to buy for him until further notice. By the end of the following month, everyone on the street knew Sharon was trying to corner the market, and shares jumped to $100. E. J. "Lucky" Baldwin held a large block of Ophir stock and, although Sharon offered to become partners with him, Baldwin held out. Expert mining engineer Phillipp Deidesheimer advised Baldwin not to sell, because there were "150,000 tons of ore standing." Finally Sharon agreed to pay Baldwin's price: $135 a share for 20,000 shares. Baldwin made $2,700,000. Sharon encouraged friends and associates to buy Ophir, confiding to them that it would go over $300 a share. More than Sharon's manipulations, the market was being driven by the wondrous developments on the 1,500- and 1,550-foot levels of the Con. Virginia.[3]

Mackay had been forced to close the Con. Virginia to outsiders, because, with the constant crush of visitors, the day shift had been able to work only half-time. At the end of October 1874, the *Gold Hill News* reported: "The future prospects of Con. Virginia are almost beyond estimate." The *Enterprise* stated that soon the mine might be working in pure silver; it concluded that the great ore body extended through the Con. Virginia, California, and Ophir to the next mine, the Union Consolidated. On December 3 Con. Virginia stock sold at $196; California, which was not yet completely open and so had not started producing, at $160; and Ophir at $118.

Mining men were speculating that, once open, California mine stocks would go to $1,000. DeQuille estimated that in the Con. Virginia, between the 1,400- and 1,500-foot levels alone, there was $116,748,000 in sight. Deidesheimer concluded publicly that the total value of the big bonanza would prove to be in the unbelievable range of $1,500,000,000. When asked how much stock he owned, he replied: "Oh, I have made a few millions; that will do me."[4]

The stock market holiday recess, from December 24 to January 2, did not stop the investment craze. Street trading continued to swell prices. By January 7, 1875, the thirty-one leading mines were valued at $262,669,940 (the sixty-five other mines brought the total market value to $300,000,000). Everyone was caught in the excitement of the moment. Flood bought 100 shares of Con. Virginia for $680,000. Reporter Dan DeQuille wrote to his sister, "in great haste," on January 6, "I will make at least $10,000 on my California and Consolidated Virginia. . . . They are getting it richer every day. . . . I am sure to make a *house* and maybe half a dozen houses." Coll Deane, later the chairman of the stock board, bought 1,000 shares of Con. Virginia for $800,000, saying it was commonly thought the price would reach $3,000 a share. Con. Virginia stock rose to $700, California to $780, and Ophir to $315.

In truth the mines were overvalued; there was not enough money on the West Coast to sustain the market's garish prices. January 7 was the apex— by the next day, doubt crept in. People began to sell, and qualms turned to widespread alarm. Brokers and street speculators panicked. Everyone rushed to sell, and the market collapsed. Much of the buying had been on margin. Ten leading mines depreciated $17,814,800 in twenty-four hours. Con. Virginia shares fell to $497, California to $240, and Ophir to $100. Over the next month, prices continued a free fall, with the Ophir sinking to $65 a share. Thousands of investors were ruined, thousands of others seriously damaged.[5]

Two notable investors not only survived, but also profited. Sharon's broker in the Ophir dealings, James Keene, had shorted the market, making a fortune. When he left for Wall Street at the end of the following year, he carried with him a reported $5,000,000 from the crash and one other "bear" action. Sharon came out in the black as well. While urging others to buy, he had sold his Ophir shares.

Lawyer and former United States minister to Japan, Charles DeLong, who invested and lost heavily in Ophir speculation, called Sharon a demon who ought to be destroyed. DeLong wrote his wife: "I fear that this whole Ophir job was put up by Sharon and that he cruelly and wickedly advised his friends to buy and to keep buying and to hold when he himself was selling and that he then as wickedly as sin contributed all in his power to smash the market thus getting his stock sold for 250 & 300 back at from 90 to 150."[6] One other investor who lost heavily in buying Ophir shares was Sharon's partner, William Ralston.

DeQuille interviewed Mackay on the Comstock shortly after the crash. Mackay said: "It is no affair of mine. I am not speculating in stocks. My business is mining. . . . I see that my men do their work properly in the mines and that all goes on as it should in the mills. I make my money here out of the ore." In an obvious swipe at Sharon, he said that he could buy substantial shares of stock, go down to San Francisco and throw them on the market at a critical time, and cause a crash to earn $500,000. "[But] what is that to me? By attending to my legitimate business here at home I take out $500,000 in one week."

Before the crash Mackay spoke with W. Wallace Austin, who worked for the *Gold Hill News.* He told the reporter that when the crash came, it would "make a greater stir than the boom has made." He counseled Austin: "If you had enough money to invest in Consolidated Virginia, there is no property in the world that would pay you better. But you must buy the stock and hold it, as I do, to get the benefit from it. . . . Speculating in stock values simply places your money at the mercy of the brokers, and they can take it from you whenever they desire."

After commenting to DeQuille that the California mine would take six months of steady work before it would reveal how rich it was, Mackay spoke of the Con. Virginia: "Some persons think that the stock has already sold for more than it is worth. The truth is that it has never yet sold for one half of its value; but all this will be seen in good time. People will see it after a while."[7]

Bank of California president Ralston was clinging to the hope that Mackay was right and that the mine's bonanza extended into the Ophir. Ralston desperately needed to be right when he told others that he had confidence in the mine.[8] He was overextended and had secretly been using bank money to cover his debts. No one afterward argued that his great expenditures were done for personal gain; his motives were ascribed to advancing the prosperity of San Francisco, California, and the West. From its inception, Bank of California monies were used for nascent enterprises that might spur development. Many of these businesses succeeded. But in 1873 four large companies defaulted on over $3,500,000 owed the bank. In that instance a loan was secured to keep the bank from failing. The loan was to be repaid by Ralston.

In the summer of 1875, the Firm was readying the Nevada Bank of San Francisco to open. The building being constructed to house it, at a cost of five hundred thousand dollars, was a massive four-story affair called the Nevada Block. Once opened, not only would the bank be in direct competition with the Bank of California, it would carry off three hundred thousand dollars by withdrawing the Firm's San Francisco accounts. Ironically, upset with Sharon over the mine tax fiasco, Mackay and Flood would be squeezing Ralston, whom they liked. Asbury Harpending, a Ralston friend, commented: "So far as the Bonanza firm was involved, its members were personal friends of Ralston, though not of William Sharon." Grant Smith later commented: "Mackay and Flood admired Ralston and had no personal differences with him. Sharon they disliked, to put it mildly."[9]

On the Comstock, Mackay had reached the pinnacle of mining success.

It was clear that the big bonanza would free him from financial worry for the rest of his life. He was abstemious in his personal habits—his drink was champagne, although not more than a glass—and he would persist in his unswerving work routine.

At the end of May the *Nevada State Journal* reported that the Mackay-Fair payroll had been fifteen thousand dollars in April and that it would reach twenty thousand dollars in the current month. It commented: "This is a small slice of the 'bonanza,' but it is very acceptable to all of us." Three months later a reporter for the *New York Tribune* discussed the bonanza kings:

> Each day I have seen three men each worth not less than twenty millions of dollars, going about quietly among the men in the common garb of the laborer, with nothing to distinguish them from the ordinary mine hand. No diamond studs, no big rings. . . . One of these gentlemen (Mackay) has spent years in foreign travel, and has mingled in polite society in other countries; another (Flood) who spends most of his time in San Francisco rules the market here. . . . the other (Fair) is accustomed to direct hundreds of men and employ millions of capital. Yet these three men, when seen about their mines, would be taken for foremen or overseers. They attend to their legitimate business as miners and earn handsome dividends for their stockholders.

In the same article the director of the U.S. Mint, with whom the reporter was traveling, remarked upon leaving the Con. Virginia mine that he had never seen so much wealth in his life.[10]

For all the riches in Virginia City, money was scarce on the rest of the West Coast. One of the consequences of the depression in the East had been a reduction in workers' wages. That decrease allowed businesses to charge less for products competing with those in the West, whose wages had remained relatively high. By August 1875 several Ralston-backed companies were in the red. The fall in the price of silver contributed to his financial woes, as did a dry year that negatively affected most agriculture production.

The wheat crop in California's Central Valley was surprisingly good, but it caused a further tightening of money, because $4 million was withdrawn from San Francisco banks to finance transporting it east. Because of its high price, over $30,500,000 in gold bullion had been shipped to New York since the first of the year ($15,000,000 more than the previous year). In August there was only $10,000,000 in free circulation in California, and the movement of funds in Ralston's bank was between $6,000,000 and $9,000,000 daily. Adding to the financial crisis for Ralston were his personal investments. Construction of the Palace Hotel, projected to cost $600,000, was running into the millions, and the Spring Valley Water Company, purchased using a complicated scheme behind which there was no cash, now required money.[11]

Ralston's hope that mining stocks, the Ophir in particular, would improve and he could hang on until money loosened seemed reasonable in early spring, when mining stocks rallied. But Mackay refused to make an estimate of ore in the Con. Virginia. This refusal, combined with his comment that the mine was undervalued by half, was taken by many as evasion and exaggeration—merely intended to boost stock prices. Raymond Rossiter's "U.S. Mineral Resources" report issued in March 1875 (from studies conducted the year before) commented that "it is impossible to estimate the amount and value of ore standing in the Virginia and California ground." To investors this seemed another equivocation.

On May 20, 1875, the *San Francisco Chronicle* ran an article titled "The Stock-Jobbing Juggernaut," which accused mine owners of manipulating the market for "lascivious" gain. It accused Sharon, Ralston, and their partner and former Bank of California president, D. O. Mills, of using newspapers they owned to tout the richness of the Ophir before levying assessments and to spread rumors depressing stock prices when they wished to buy. It condemned practices that caused Ophir stock value to drop from $315 in January to $36, accusing the owners of "conspiracy" and "robbery." The editorial concluded: "Dealing in mines under honest and honorable management is a hazardous business. Dealing in stocks under dishonest and dishonorable management will result in inevitable loss." Combined with the vacillating Comstock reports, this attack dampened mining investments, resulting in a steady decline in stock prices.[12]

In mid-August the market was feverish; the leading stocks, which had gained somewhat at the beginning of the month, again sagged. Flood kept in contact with Mackay. He also met with an assistant, N. K. Masten, who later became president of the First National Bank of San Francisco, and his broker, Colonel Edward Eyre. (Eyre, who was partial to corn juice although it reportedly never interfered with business, earned $5 million in commissions from the Bonanza Kings.)

In mid-August people connected with the Bank of California began transferring property to other names. Taking note of this, on August 20 Flood sent a note by messenger to Ralston. The messenger reported that Ralston read the note, flushed deeply, and said, "Well, you go back and tell Flood that I'll send him back selling rum over the Auction Lunch counter!"

The messenger recounted that Flood replied: "You go right back and tell Ralston that *Mr.* Flood says that, in a short time, he will be able to sell rum over the counter of the Bank of California." Fifteen minutes later a pale Ralston was in Flood's office. When Ralston hurried back out, Masten rushed in with a warning: "Flood, draw all of your money from the Bank of California! It's going to fail!" Flood's original message had told Ralston that the Firm wished to withdraw its $300,000 in coin. At Ralston's request he had accepted $100,000, although continuing to demand coin, with the balance to be paid in two installments on successive days.[13]

People knew the battle between the Bank of California crowd and the Bonanza Kings was coming to a head. Ralston told bank director Thomas Bell that Flood and O'Brien "were locking up all the coin and making money very scarce." Rumors spread that the Bonanza men had withdrawn millions from the Bank of California. Ralston and Flood gave interviews to the newspapers, each describing their relationship as pleasant and denying the withdrawals.[14] Ralston began selling his properties, and San Francisco newspapers deduced that the bank needed funds. Restlessness in financial circles increased.

On Wednesday, August 25, there was a noticeable drop in all stocks. Con. Virginia lost $45 and California $6. That night the Bank of California directors met and the cashier announced that instead of the $2 million thought to be in the bank vault, there was only $500,000. The rest was accounted for with tags signed by Ralston.

The next morning, August 26, Sharon and D. O. Mills met with Flood, O'Brien, and their representatives. Sharon offered to turn over the bank, with its $5 million capital stock, $1.5 million reserve fund, and another million on top of that, if Flood & O'Brien would assume the bank's liabilities and hold the directors and stockholders harmless. The Firm decided against assuming the liability.[15]

When the stock-exchange board met at 11 that morning, Sharon torpedoed the market. His broker began selling Con. Virginia and continued until no one else would buy. The great mine had opened at 267½, already thought to be a cut-rate price, and fell steadily to 240. Sharon's order to sell had been unlimited, and California stock dropped from 56½ to 48. A rumor quickly spread that Sharon was trying to get cash to prop up the Bank of California. (He was not; he deposited all he made in a personal account in Wells Fargo.)

At 2 PM a crowd surrounded the paying counter of the Bank of California. By 2:15 the crowd had swelled to the bank's iron doors. Within minutes the sidewalk was teeming with people who could not get inside. At 2:35, twenty-five minutes before closing time, the great doors swung shut. Unable to pay out any more cash, the Bank of California had suspended.[16]

At 3 PM word of the Bank of California's suspension had been received on the Comstock. The Virginia City branch closed, and a statement was issued that although it would remain closed until further notice, the first concern was to protect depositors. The streets filled, and an excited crowd gathered outside the branch office. Amid the chaos Mackay took action, organizing a savings bank so monies could be protected. The directors included himself, Fair, their timber foreman J. B. Hereford, attorney Charles DeLong, a state senator, and nine other leading citizens. As crowds milled about, a detachment of the National Guard was placed on alert at the armory. They remained there through the night, and the brigadier general of the Nevada State Militia remained inside the bank. The precautions were unnecessary, as by midnight the streets emptied, but excitement rose again the next morning, as word came that San Francisco's money market had failed. The *Gold Hill News* commented that the suspension was unlike any other: "[It] was so totally unexpected and incredible, that if the heavens had fallen our people could not have been worse confounded."[17]

In the morning of August 27, 1875, Ralston signed over his possessions to Sharon and resigned as bank president. In the afternoon he swam out into San Francisco Bay and drowned. All San Francisco grieved for its greatest citizen. The West's economic structure was shaken to its foundation.

The news of the death of William Ralston, received the evening of the 27th, created a tempest on the Comstock. The streets were filled, and rumors spread: broker James Keene had shot Sharon; the offices of the *Bulletin* and *Call,* newspapers that had been criticizing Ralston, had been attacked by a mob.[18] The rumors, which were untrue, did not reach Mackay. He had left for San Francisco first thing the morning of the 27th. By the time he arrived,

Ralston was dead, his associates in shock. Sharon met Mackay at the ferry. As they drove uptown in a hack, Mackay referred to Ralston's bankruptcy, asking: "Did you do everything you could to save him?"

Sharon answered: "I was afraid for a time that we would have to." In telling the story, Mackay referred to Sharon as "that son of a —."[19]

In San Francisco and throughout the state, runs were made on other banks, resulting in the temporary closing of several. The three major banks in Los Angeles shut their doors. Several institutions issued statements that although their assets were sufficient to meet all liabilities, they would not open the next morning. The San Francisco Stock Exchange and the newly formed Pacific Stock Exchange also suspended operations.[20]

Public castigation of those who had opposed the fallen hero began with the newspapers that had campaigned against Ralston. The night of his death a large crowd gathered outside the *Call* and *Bulletin* offices, but because of a large police presence (and contrary to rumors), no violence had occurred. When the papers continued their attacks on the Ralston political machine and his political allies after his death, it spurred a public backlash and a substantial loss of advertising and circulation for the *Call* and *Bulletin*.

The Bonanza Firm came under fire for Flood's actions at the time of Ralston's tribulations. On August 27 the *Chronicle* charged that "instrumentalities," commonly understood to mean Flood and the Firm, had locked up a large sum of the city's money, causing the failure. On the 28th Flood gave the paper an interview but did not respond to the reporter's question regarding why he had insisted on coin when withdrawing a great sum the week before. On the 29th the paper reported that a throng gathered in front of the *Chronicle* office. One crowd member voiced the opinion shared by others: "Flood & O'Brien have precipitated this calamity in San Francisco, and every sensible man knows it. They have got what they wanted from the Bank of California, or they will get it, and in doing so they have run the risk of plunging the entire city into financial ruin."[21]

A reporter from the *Alta* had an audience with Ralston just hours before his death. The reporter said to Ralston: "Mr. Flood states that he doubts the ability of your bank to pay its depositors in full. What is your idea of the matter?" Ralston replied that depositors would be paid "to the last cent" and remarked: "Must say that the doubt was rather unkind." When asked if one group had locked up the city's coin, Ralston was evasive. He stated that the parties involved had disputed the charge and he would leave it at that. When specifically asked if Flood and O'Brien had withdrawn $1.8 million on the previous day, he said: "It is all nonsense, all balderdash. . . . So far as our relations with those gentlemen are concerned, they are of a character perfectly pleasant and agreeable, nothing unpleasant in any way, shape or form; nor has there been any such words as were reported—none whatever."[22]

The *Alta* interview concluded with the reporter's asking if Ralston believed that assistance from his friends would have staved off the difficulty. Ralston exploded: "Damn it, sir, that is rather a sour subject which I do not care to discuss. Good afternoon." In the end a preponderance of pundits concluded what the reporter inferred: that, other than the victim himself, his associates—in particular his partner Sharon, had been most responsible for Ralston's financial failure. By August 30 the *Chronicle* had come to that conclusion: "Mr. Ralston in the hour of his extremity, and at the moment when he most needed encouragement, was abandoned by those who should have felt their fortunes linked to his own with bonds strong as gratitude and humanity could bind them."[23]

In the aftermath of the tragedy, Sharon realized that because he was Ralston's partner, Ralston's liabilities had become his. He rallied the wealthiest San Franciscans and, in one of the great feats in nineteenth-century banking, reestablished the bank within weeks of its failure. It reopened on the same day as the gala opening of the eight-hundred-room Palace Hotel, originally Ralston's cherished project, now solely owned by Sharon. The newspapers hailed the twin feats with ebullient fanfare, completely overshadowing the Bonanza Firm, which began business at the Nevada Bank of San Francisco the same day.

The Comstock had problems of its own in the days before the collapse of the Bank of California. Alf Doten thought that perhaps one third of its inhabitants were suffering from cholera, commenting in his diary that children were dying off at a fearful rate. On the morning of August 13, a fire began in the boiler room of the Imperial mine works. It was only through firefighters' great exertions that the structure was saved from complete destruction.[24]

On September 3 another big fire erupted. This time the Odd Fellows lodge and the Metropolitan Livery stables were completely destroyed. But on October 2 the Bank of California reopened its Virginia City and Gold Hill branches without a run. The Gold Hill agency displayed one hundred thousand dollars in its trays, and the citizens reaffirmed their confidence in the institution by making more deposits than withdrawals. That evening there were speeches amid a celebration that included bands, bonfires, and a torchlight parade. Two weeks later, when Civil War hero General Phil Sheridan and his brother Colonel Mike Sheridan and their wives arrived on the train from San Francisco via Lake Tahoe, the town was draped with flags, and a grand cannon, whistle, and bell reception brought everyone out. Normalcy had returned. The populace again frequented the shops and restaurants—described by a *New York Tribune* writer as world class.[25]

On A Street, above the shops and restaurants on B and C, a woman named Kate Shay, known commonly as "Crazy Kate," kept a lodging house said to be of low repute. On October 25 those at the house stayed up carousing through the night. Early on the morning of the 26th, a fire broke out. A few minutes after 6 AM, an alarm was sounded, and in short order the hand engine of Fire Company No. 4 and a fireman with the eight-gallon Babcock fire extinguisher strapped to his back were on the scene. But a strong wind, blowing from the west, carried cinders ahead of it. Church bells and mine whistles began clanging and shrieking.

Twenty minutes after the alarm was sounded, forty buildings were ablaze. The heat and wind combined to cause whirlwinds of flames. Brick buildings were incinerated. The *Gold Hill News* described the north end of town as a "seething sea of flames." The *Territorial Enterprise* called the blaze "a breath of hell" melting the town.[26]

Mackay and Fair directed work in fighting the fire at the Bonanza mines. They had the cages of the Con. Virginia piled with sand and ore and lowered a few feet into the shafts, which were then covered with a heavy platform and piled with dirt to prevent the fire from moving underground. The wood floors of the hoisting works were covered with three or four feet of sand. Miners were dispatched to dynamite the houses on the west side of D Street in an attempt to stop the blaze. In the midst of the work, an old woman came to Mackay telling him St. Mary's Church was on fire. C. C. Goodwin reported that Mackay replied: "Damn the church, we can build another if we keep the fire from going down these shafts!"[27]

The fire reached its peak at the Con. Virginia and Ophir works. Despite the dynamiting of outbuildings at the Ophir, sheets of flames swept down and all hands fled to save themselves. The next day the *Territorial Enterprise* reported: "About eleven o'clock on Tuesday, out of the crater of the conflagration there emerged a man haggard with fatigue, begrimed with dust, powder-smoke and the smoke of the fire. He looked like a laborer just ready to drop from exhaustion. It was John Mackey [*sic*]. When it was determined to blow up the houses below G Street to save, if possible, the C. and C. hoisting works and the new reduction works below, he, with his old miner instinct and miner's knowledge, led the charge."

Mackay's efforts were to little avail. The Con. Virginia's ore house, hoisting works, new sixty-stamp mill, refining and assaying houses, shops, and offices, as well as the California mine's unfinished stamp mill, were incinerated. A million feet of stored lumber and thousands of cords of firewood burned with such intensity that nearby railroad cars' wheels melted. The heavy timbers of the Ophir's gallows frame collapsed, breaking through the earth-covered platform into the shaft. Flames began to work their way down it.[28]

The ongoing danger was that the fire might gain the timbers that lined its shafts and tunnels underground. Once a mine's framework burned, damage was irreparable. Underground fires were known to burn for years, causing stopes to cave in, mixing waste rock and ore. Sam Curtis, the superintendent at the Ophir, organized lines of men who "stood in the midst of the burning mass, passing water-buckets from hand to hand." The work was continuous by the bucket brigade until midnight and through the night by fire engines, as they were freed from other duties.[29]

By noon the heart of the town was a smoldering ruin. Miraculously, there had been no loss of life, but all principal hotels and rooming houses had burned. The *Gold Hill News* announced that fully ten thousand people were homeless. "In all directions and on all the streets the people were seen lugging along trunks, articles of furniture and bundles of bedding and clothing." Editor Doten commented that over the divide, Gold Hill had opened its houses to the evacuees. Still, thousands of people were left to sleep out in the sagebrush. Mackay's house, insured for ten thousand dollars although worth twenty-five thousand dollars, was listed as one of the principal buildings among the blocks of destroyed houses and businesses. When asked about his losses, Mackay said they were not worth mentioning, that others were more in need. The Bank of California's loss was estimated at thirty-five thousand dollars, while the Washington saloon and Mallon brothers' provision store each lost fifty thousand dollars. The Bank of California was fully insured. Few others were. St. Mary's and the other major churches had been lost. So too were all houses on both sides of D Street, the *San Francisco Chronicle* commenting that they were "occupied by a class of population which will be no great loss to the city."[30]

Help began to arrive on the 27th, the day after the fire. Two train carloads of food arrived, and a cable from Carson City announced that thirty thousand dollars had been collected from its citizens. Another from San Francisco said brokers had donated six thousand dollars.

At the disaster site there was mutual assistance. Frank F. Osbiston, superintendent of the Savage mine, whose lower levels were shut down by a heavy body of water, loaned that mine's pumping works to help keep the Con. Virginia and California mines free of water. Fair, who superintended the Chollar-Potasi, also moved its pump to assist. The Phil. Sheridan Mining Company likewise furnished Curtis at the Ophir mine with a donkey pump. The Virginia & Truckee Railroad provided free transportation to people forced to leave Virginia City to find winter refuge in California. Alf Doten, now the editor of the *Gold Hill News,* allowed his competitors, the *Virginia Evening Chronicle* and the *Territorial Enterprise,* whose offices had been destroyed, to use his offices to publish their papers. (The *Enterprise* utilized the cramped space for over two weeks, seventeen issues, until it moved to the basement of the Odd Fellows building.) The "saloon men" were said to have supplied sufferers with whiskey despite the fire, smoke, and dust. They were praised for having neither watered down the whiskey nor raised their prices.[31]

The *San Francisco Chronicle* of the 27th tried to assuage public fears to avert a market crash, telling its readers that the first news of a catastrophe is always exaggerated and that the mines and bullion were still there. Although the mills and hoisting works may have been lost, and production would certainly be suspended temporarily, it reassured the public that the mines had not been destroyed.[32]

The following day the *Chronicle*'s analysis was thrown into question when smoke began issuing through the seams of the platform covering the Con. Virginia. At the Gould & Curry, a quarter mile away and connected by tunnel, the smell of gas was emitted "as strong as from the gas works." These were almost irrefutable evidence that there was a fire inside the Con. Virginia. Fair issued a statement that he was certain there was no fire inside it, and that the small amount of smoke coming from it was a result of the deep connecting tunnel with the Ophir. Curtis denied his mine was on fire any more than twenty-five feet below the surface and continued pouring water down the shaft.[33]

Public confidence was not allayed. Prices on the market plunged. In an hour Con. Virginia stocks fell to $100 a share, a loss in market value of $10,800,000. California's 540,000 shares dropped $10 each. The decline of other stocks was even more precipitous.[34]

On October 28 an interview with Flood was printed in the *San Francisco Chronicle*. He could provide little definitive information, except to say that the Morgan and Brunswick Mills—both belonging to William Sharon, with a total capacity equal to those burned—were available, as were other smaller mills. The hoisting works of two mines had also been offered for sale to the Firm, though no decisions had been made. Reconstruction of the lost buildings would begin immediately, but ore production might be delayed thirty, sixty, or more days.[35]

On the 30th an article on Mackay appeared in the *Chronicle*. The reporter was amused when Mackay was pointed out to a gentleman asking to see him. Mackay was talking with a group of miners, and the man would not believe that the owner of the richest mine in the world was dressed in miner's clothes, standing amid the twisted metal and debris of the fire. When the reporter talked with Mackay, it was at Fair's house, described as a "palatial residence," where Mackay was staying. Mackay said he was certain the Con. Virginia would be hoisting ore within thirty days. He had inspected it, the California, the Ophir, and the Best & Belcher, spending three hours underground that afternoon, and saw no sign of damage. "There is not a single timber or wedge, so far as I could see, out of place. . . . Our mines are all right in every respect, and if the insurance men will pay off without delay all these poor people around us, so that they may go on and build, Virginia will soon be herself again." The reporter commented: "This sudden turn of the conversation from the big bonanza to the condition of the citizens at large, so characteristic of Mackay, ended the dialogue, and the reporter, after accepting a little of the purest Bourbon and a fine Havana, took his departure."[36]

The reporter would have been more impressed with Mackay's concern for the town's citizens if he had been following his movements after the inferno. Although haggard and fatigued from fighting the fire far into the night, the next morning Mackay visited Father Patrick Manogue, the pastor of St. Mary's Church. The cleric was an almost mythic figure: six foot three, 250 pounds, nicknamed by the Irish immigrants "Soggarth Aroon," a beloved priest. His parishioners numbered more than those of any other religion in the territory, and he was often called on to mediate affairs outside his church.

Before the fire St. Mary's Church featured a marble alter from Carrara, Italy, and Stations of the Cross painted in Florence, each procured at Mackay's expense. Mackay now offered not only to rebuild the church, but also to help all who required it. He told Father Manogue to use Mackay's resources as long as necessary.[37]

The fact that Mackay chose Father Manogue to distribute his donations and contributed to Catholic charities throughout his life highlights an interesting dichotomy. He was raised as a Catholic, and his sister served her life in a convent. He was married in the Catholic Church, his children were raised as Catholics and, although he did not attend Mass, on his deathbed he received Last Rites from a priest. Curiously, he was a Mason as well, even though the society's relationship with the Catholic Church was mutually exclusive. In Virginia City he became affiliated with the Masons' Escurial Lodge, and in 1873 he became one of the five life members of the Grand Lodge of Nevada. He rarely spoke of his membership in the Masons, perhaps out of concern for Louise, a practicing Catholic. Regardless of his affiliation, Virginia City newsman C. C. Goodwin said that over the next three months, Mackay gave Father Manogue some $150,000.[38]

As soon as the fire burned out, with the ruins smoldering, the work of rebuilding began. Mackay's statement to the reporter that there was not a single timber out of place underground was true in his own mines but not in the Ophir. Its hoisting-shaft timbers had not been extinguished until the fire reached a depth of four hundred feet, and the structure needed to be replaced. Curtis's reconstruction efforts typified the building that went on throughout the city. The annual report of the U.S. Commissioner of Mining Statistics for 1877 stated: "On the day after the fire competent men were dispatched to the lumber-yards of Carson and Dutch Flat, Cal., to procure and ship timbers; machinery was telegraphed for . . . work was prosecuted without cessation, supplies hauled a considerable distance on account of destruction of railroad tunnel and bridges, the works rebuilt and work through shaft resumed November 25, being inside of thirty days from time of destruction."[39]

Work began early in the morning and continued late into the night. The week after the fire, a tornado blew down many of the walls that had just gone up. The wreckage was cleared and work was restarted. At one point Goodwin mentioned the famine in Ireland to Mackay, who responded: "There are a good many poor people right here, but you may thank God that none of them are either cold or hungry." Goodwin did not find out until later that Mackay was supplying the money that clothed and fed the fire victims.[40]

After the fire the Virginia & Truckee Railroad had to replace its depot, trestles, bridges, switches, track, and the timbers of a tunnel within the city. When this work was completed, trains began running day and night—as many as forty-five in a twenty-four-hour period. Sixty days after the fire, businesses lined the main streets of town and houses filled the adjoining lots. Mackay had moved into the house built for the superintendent of the Gould & Curry mine, where he would live until moving from the Comstock.[41]

He missed his prediction of thirty days to restart production in the Con. Virginia by two weeks. Its rebuilt hoisting works began pulling ore to the surface on December 13, averaging six hundred tons a day thereafter. The new works were enlarged, including a two-story building with facilities for melting and assaying five million dollars' worth of bullion a month. Mackay purchased ground around the hoisting works to increase space for storing wood and mine materials. This isolated the works, so there was little danger of fire unless it began on their premises. The works were surrounded with hydrants and water under heavy pressure to provide security. The Virginia & Truckee railroad tracks ran around the works on three sides, so that supplies were delivered on the uphill side and ore was shipped out from the downhill.[42]

At the end of 1875 the buildings for the c & c shaft, constructed for use by both the Con. Virginia and the California mines, were completed. They included a blacksmith shop and machine shop, each divided in half for use by the two mines, and two large carpentry shops. Shortly after their completion, adjacent to the shaft, the new California stamp mill was finished. It was able to crush 280 tons of ore per day (older mills crushed from 50 to 180 tons per day). The California pan mill, five hundred feet east of the mill, had survived the fire. Crushed ore was conducted to the pan mill through a pipe four inches in diameter. Timber, wood, and lumber costs for the year were $285,437, and construction outlay was $96,935. Despite the unforeseen expenditures, the Con. Virginia continued to pay monthly dividends at $10 per share.

Fair was now superintending the Con. Virginia and California mines, as well as the Best & Belcher, the Gould & Curry, and the Hale & Norcross.[43] If one were charitable, it might be said that his misrepresentations in the 1874 and 1875 annual reports of the Con. Virginia were due to the stress of overwork in managing so many properties, or perhaps they were due to confusion with all the rebuilding after the fire. However, Fair's boasting and bullying gave no cause to look at his actions with benevolence, and common opinion was that he exaggerated or lied in the reports in an attempt to bull the market and raise the value of Con. Virginia and California stocks.

The 1874 report said that the northern 400 feet of the mine's drift ran through ore assaying from $300 to $800 a ton. If the assays were so high, investors wondered, why was the average yield per ton in 1875 only $93 (even though this was a record yield at the time, broken the following year when Con. Virginia ore averaged the all-time record of $114)? In the 1875 annual report, Fair magnified the amounts of ore at the 1,400- through 1,550-foot levels and lied about the placement of a double winze sunk from the 1,550-foot level to a depth of 147 feet through very high-grade ore. This faulty data proved, he concluded, "the continuity at these lower depths of the same ore body which exists on the level above, with an appreciation in the quality of the ore." The difference in the placement of the winze made it appear that there was an additional 160 feet of ore and doubled the length of the strike below 1,550 feet at the south end. The distortions in the annual reports left the Bonanza Firm susceptible to the attacks that soon came.[44]

CHAPTER 8

Responses

ONE OF THE BONANZA FIRM'S flumes was fifteen miles long, descending nearly two thousand feet from the top of its course in the Sierra to its terminus above Washoe Valley. Two-inch planks arranged in a V shape atop trestle works formed the flume, which, with a steady stream of water, carried lengths of timber down the mountain. The steep grade prevented logjams. In one place the works stood seventy feet above the ground.

Daredevils sometimes rode down the flume for excitement. On one occasion a reporter, traveling with Fair and Flood to peruse the Firm's logging operation, accepted their challenge to ride it with them. He reasoned that if the multimillionaires were going to risk their lives, he could afford to risk his.

The two "boats" to be used were sixteen feet long and V shaped to match the flume. Each had an end board at the rear of the craft, so the current would propel it along. Narrow boards were placed in each boat to serve as seats. A whiskey-red-faced carpenter, the only volunteer from a fifty-man crew that was watching, was to ride in the front of the first boat, then Fair, then the reporter. In the second would go Flood and J. B. Hereford, the superintendent of the lumber and flume company. As the boats touched the water, the men awkwardly boarded and were hurled forward. They could not stop; they could not lessen the speed. At the steepest grades, water cascaded upon them so furiously that it was impossible to see ahead. At lesser grades, the view—twenty to seventy feet above the earth—was terrifying. "When the water would enable me to look ahead," the correspondent reported, "[the trestle appeared] so small and narrow, and apparently so fragile, that I could only compare it to a chalk-mark, upon which, high in the air, I was running at a rate unknown upon railroads." Objects he tried to fix upon on the hillsides were gone before he could focus. The mountains passed "like visions and shadows."

When the lead boat struck a limb caught in the flume, all were thrown face forward, with the carpenter tossed out ahead of the craft. The powerful Fair grabbed the man by the scruff of the neck and pulled him back into the boat. Fair's fingers were crushed between the boat and the flume in the effort. At the steepest declination, estimated to be forty-five degrees, the boat seemed to fall, somehow managing to stay in the track but so taking the reporter's breath that he thought he would suffocate. At the terminus the soaked riders disembarked "more dead than alive."

The next day the reporter said neither Flood nor Fair left his bed.[1] It was not surprising. Fair was forty-four years old, Flood forty-nine. Plucking the carpenter from the flume demonstrated Fair's uncommon strength and athleticism. In the early days it was said that he was always ready for any kind of fun. The ride illustrated that he had not abandoned that attitude, as well as exhibiting his daring.

But despite Fair's winning qualities, from the time the Firm came to public notice he was compared to his partner Mackay, and in almost every instance he came up short. It was Mackay who was often mentioned as U.S. Senate timbre, never Fair—even in the years after he spent a small fortune to procure the Nevada seat. Mackay's motives were generally assumed to be the highest; Fair's never were. Mackay was the hero of the common man. On the Comstock Fair lived in a mansion, Mackay in rooms at the Gould & Curry office. Both were familiar figures on Virginia City streets, but it was pointed out that Fair was driven about by a coachman in a carriage drawn by a team of bays, while the unassuming Mackay drove himself in "a shabby buggy."

When kids sledded down the town's steep cross streets ending at the Con. Virginia works, if Mackay's buggy was outside the boys waited. They knew he would allow them to tie their sleds on and would tow them back. If Fair's carriage was there, they started the long, uphill hike back.[2]

There are stories of Mackay's buying boots for barefoot boys on the Comstock, and an everyday largesse was described in a Virginia City reminiscence: "Each [new] performance in Piper's Opera House meant forty or fifty penniless small boys hoping for miracles as they stood outside the entrance." When Mackay happened by, he would simply ask the proprietor how much for the bunch—generally about twenty dollars—and buy the tickets. "We would enjoy the show, and what is more we would think better of all mankind because John Mackay had remembered that he was once a boy."

Another who grew up on the Comstock commented: "Every boy wanted to grow up to be a man like Mr. Mackay." And yet another: "The sight of Mr. Mackay would set us small boys all aglow. We thought he was one of the wonders of the world."[3]

Historian Ronald James points out that Mackay and Fair represented starkly contrasting saint and sinner in Comstock lore. Mackay was seen as humble, honest, and charitable. The miserly Fair, "in an ironic play on his name was popularly regarded as unfair. Fair became the ultimate sinner in local legend, counterbalancing the reputation of Mackay and illustrating how wealth could corrupt a man with flaws."[4]

Fair, in giving others investment tips, was generally thought to have ulterior motives. An old friend, Colonel N. H. A. Mason, the cattle king of Mason Valley, disabused the notion: "I have heard much about Fair being slippery, but he never deceived me. I have asked him for pointers of stocks, but he used to say, 'Mason you're a water man and a cattle man, and you had best stick to that.'" But an often-repeated story told of Fair's letting false information "slip" to induce his wife and her friends to invest in the faltering Gould & Curry mine. When they bought the stock, the market rose and he unloaded his shares. The mine's state of affairs then became public knowledge, and Gould & Curry value collapsed. Fair returned his distraught wife's lost investment, chiding her on her lack of investment skills.[5]

Mackay resented the onus of others' expecting him to assist them in investing. Dr. A. M. Cole had been a stockholder with Mackay in the Milton mine in 1862. In those days Mackay could be found evenings outside Cole's drugstore or inside it on summer evenings. They remained fast friends through the years, and an acquaintance mentioned to Cole that he ought to be rich on Mackay's tips. Cole replied: "You don't know the man. He knows I have a thousand shares of Con. Virginia, and that I'd like to know whether to hold or sell, but neither of us says a word about it." Another anecdote reveals Mackay's consideration toward his friend: a miner came to Cole, informing him that rich black ore had been struck in the Hale & Norcross. The next night the miner returned, saying the black stuff turned out to be base metal of no value. Cole had bought the stock and now sought out Mackay. "Did you get caught, too?" asked Mackay. "I bought 25,000 shares but the damned stuff is worthless. I am going to sell my stock but I'll let you get out first."[6]

On March 6, 1876, the *San Francisco Bulletin* published a feature article describing the members of the Firm. Mackay, it said, could be found at nearly any hour of the day, trousers tucked in boots, felt hat pulled down over clear, blue eyes, moving through the snow between mines. "He has none of the airs of a monarch. If you happen to know him he will say a few pleasant words and bid you good morning. His demeanor is invariably quiet and modest, and the most jealous eye can detect no bluster in it." It concluded with the almost perfunctory mention, "He is talked of some for the United States Senate."

Regarding Mackay's partner, the *Bulletin* correspondent said: "[Fair] is a very different sort of fellow." Full of bonhomie, "(the people hereabouts call it blarney)," good looking and a good talker, "but [he] lays back in state in his buggy and mentally puts up jobs on the stock market." The article concluded: "Very sly is 'Slippery Jim,' as his admirers call him."

Fair coped with the snide comments by utilizing defense mechanisms of subtle and not-so-subtle fashion. He worked around the clock, taking on all responsibilities great and small. In a week in July 1875, a month he supervised mining ore worth $1,650,000 in the Con. Virginia, he took time to rebuke a merchant who overcharged the company by $20 on eight-inch leather belting. He met twice daily with mine supervisors, insisting: "Whenever anything went wrong people looked to me to put it right again." He spent only a few hours a night in bed, fearful of what might occur while he slept. And when things went right, he took sole credit. He was sociable and loquacious, especially when plaudits might be assigned.

Whenever important visitors or reporters came to see the mines, Fair dropped his other duties to act the host. Mrs. Frank Leslie described her visit to the Con. Virginia in 1876 when Mackay was back East. She spoke of Flood and O'Brien in San Francisco and her guide at the mine—James Fair, who apparently was her source of information regarding the Firm. Her report begins: "This mine is principally owned by Messrs. Flood & O'Brien, and Mr. Fair. The two former kept a small drinking saloon in San Francisco," and, she explained, when an inebriated customer leaked information on the "lode," they were lucky to have been there. Fair, equally rich, lived "on the spot," and ran the entire operation. Fair appears to have been circumspect in describing the company to Mrs. Leslie, omitting mention of the primary owner, Mackay. Generally, when interviewed in those years, Fair would mention all three of his partners, describing how hardworking and conscientious they were and commenting that it was only through inexperience or bad judgment that they got the Firm into trouble. Years later, after Fair had fallen out with his partners, he was more blatant about claiming the glory, saying: "Those lads would be in overalls but for me."[7]

After the Bank of California's collapse, Flood emerged as the preeminent force on the stock exchange. The former commissioner of mines, J. Ross Browne, wrote several letters regarding mining stocks in the fall of 1875, blaming a depression of prices on Flood and O'Brien. He said that it was rumored they would not permit stocks to advance until they had all they wanted. "Everybody says Flood and O'Brien are keeping things down; and there is a good deal of ill feeling against them."[8] There was another factor affecting the market at that time: the number of stock investors had dwindled. The failure of the bank and the death of Ralston had weakened confidence in the West's institutions, pointing up the fact that in stock gambling, even the most powerful insider could be brought down. Consequently, into December 1875 the market remained soft. Then, owing to the Con. Virginia's continuing success, stock prices surged. The mine's shares rose to the dizzying height of $400, pulling all others up with them.[9]

The centennial year of 1876 began with the Mackay family's deciding that Louise would move the family to New York City, while John would continue tending business on the West Coast. A great national Centennial Exposition was to be held celebrating the 100th anniversary of the Declaration of Independence that summer in Philadelphia. John would escort Louise and the children cross-country before representing the Firm at the exhibition as one of nearly thirty-one thousand exhibitors from thirty-eight states and fifty countries. The Firm was in the midst of deciding on a display to represent their mines. In January they announced that during the month of May, they would mine $10,000,000 in ore from the Con. Virginia and California mines and mill it for show at the exposition.

The proposed exhibition was debated on both sides of the Sierra. Could it be done, and was it worth attempting? The *San Francisco Chronicle* published its conclusion that it could. If the mines produced an average gross yield of $250 per ton, which the Con. Virginia had been doing, and all available shafts were utilized, 2,850 tons might be brought to the surface daily. Sharon and water-company president and lumber magnate William S. Hobart had offered additional mills. When their capacity was combined with the Firm's, 2,250 tons could be crushed a day. That meant a thirty-day run would produce $16,875,000 worth of ore. But by March, events in San Francisco forced Flood and Mackay to abandon the idea.[10]

For many weeks early in 1876, Fair was not on the Comstock. Grant Smith believed he had gone to the East. Oscar Lewis explained that Fair worked so hard he sometimes kept himself going with "copious draughts of a brandy-bottle" or, when that failed, unwanted vacations. He writes that Fair was ill enough that spring to spend two full months in California.[11] Wherever Fair was, early in 1876, Mackay was superintending the mines when they came under a furious bear-market attack.

In the stock market the term "bear" refers to those who sell shares they do not own, to be transferred at a later date. Bears are selling for a fall, betting that the value will drop and they will be able to pick up the necessary shares at a lesser rate than they agreed to sell them for. If successful, such a transaction affords them their profit on account day. Once a bear agrees to sell stocks he does not own, he needs to drive the market down or suffer the account-day loss. The optimum situation for a bear is one of high stock prices and a jittery public that can be frightened into panic selling.

The peaking stock prices at the end of 1875 continued into 1876, creating an opportune moment for clever James Keene to act the bear. Born in London, educated in Lincolnshire and then Dublin, Keene had moved to the United States when he was fourteen. In Nevada as a young man, he ran a string of mules hauling goods to the mines before becoming a stock dealer and eventually president of the San Francisco Stock Exchange. Now in the opening quarter of 1876, leading other powerful brokers, including an early-day Stock and Exchange Board member C. W. Bonynge, he began an all-out assault to break the price of the stocks. Flood was outraged. It was his first experience of the kind, and he struggled to find a strategy to defend the Firm's interests.[12]

Keene was telling everyone who would listen to sell. In the high-stakes feud that developed between Flood and Keene, publicity was accorded Eilley Orum Bowers, the "Washoe Seeress." Mrs. Bowers, the widow who, with her husband, had made a million dollars in the Gold Hill mines and frittered it away on foolish extravagances, had turned to telling fortunes: divining new strikes for willing ears.[13]

In April 1876 Flood proclaimed that when Mrs. Bowers had prophesied bonanzas in the Utah and Savage mines, Keene listened and began buying stocks. Flood said he sold Keene as much as he could buy, but that then it came time to assess the mines. The Savage, the Chollar, and the Hale & Norcross mines were sinking a joint shaft.[14] Requiring money for it and additional machinery, a two-dollar assessment was levied on Savage stock. Flood said: "Keene fancied we wanted to freeze him out, and hoping to get even he went for the bonanzas." Keene, Flood alleged, would "burst [Con. Virginia] stocks down to fifty dollars per share" if he could.

Keene called Flood's statement "utterly without foundation, and unequivocally false." He denied ever having met Mrs. Bowers or having had communication with her. He said he bought or sold stocks and advised others solely on the evaluation of the mines. He believed all indications were that the Con. Virginia ore was "getting poorer" with depth and might in fact pinch out with no ore below the 1,640-foot level. He said that Flood was upset because he wanted to sell his own shares but was compelled to reinvest to sustain the market until investors began buying.[15]

To quell the alarm, Mackay ordered that richer veins in the Con. Virginia be tapped. In March the mine produced a record $3,634,218. This caused the *Nevada State Journal* to comment: "This immense yield for one mine, for a single month, in the most inclement season of the year, we believe to be without a parallel in the history of silver mining, and must without dispute place the Consolidated Virginia at the head of all the great silver mines of the world."

Instead of quieting the bears, the production and hype pushed the San Francisco brokers to more virulent assaults. Keene asserted that the mine had been failing for some time and the large amounts of ore removed over the previous three months meant the stocks would depreciate if a new body of ore was not found below the 1,550-foot level. Flood, making every effort to prop up prices, insisted that he had sold none of the Firm's shares and advised the public to hold on to their stock and buy as much more as they could afford. He exhibited a certificate for sixty thousand shares at his club to show associates he had not been selling. He invited the public to visit the mine to see for themselves the magnitude of ore. Few availed themselves of the opportunity; cracks had developed in the foundation of the market. Keene's efforts, lasting through the summer, eroded prices. Although the Con. Virginia would yield $15,315,613 and pay $12,960,000 in dividends in 1876, shares of stock fell from $440 at the end of February to $240 on July 13.[16]

As the market war raged, Fair returned to Nevada and Mackay prepared to take his family to the East. Ronald James points out that in popular opinion on the Comstock, Louise Mackay and her good friend Theresa Fair were the opposites of their husbands. Theresa Fair took part in local charities and was regarded as generous to her husband's churlishness. Louise Mackay, in contrast, was held in disdain after 1876, when she abandoned the West Coast for the high society of the East. But Louise left a generous legacy when she moved. Jacob L. Van Bokkelen, provost marshal of Virginia City's National Guard, who had been killed when overstocked nitroglycerin that he imported for the mines exploded in his home in 1873, left his pastoral Beer Garden on the eastern extremity of Virginia City at Six Mile Canyon.[17] Louise had John purchase the land to build St. Mary Louise Hospital, a four-story building to be run by the Daughters of Charity. The Catholic facility had spacious, well-lit rooms with marble-topped washstands, hot and cold running water, and a reading room with a large number of books, magazines, and newspapers. The chapel on the third floor featured an attractive altar. The top floor had a small room with barred windows for the mentally ill and comfortable quarters for the staff.[18]

Louise wanted to leave San Francisco for New York City, because it was more established and rooted in tradition. Her O'Farrell Street house was furnished in the popular ornate French style, but it was not on Nob Hill, where Stanford, Mark Hopkins, and Charles Crocker were building their incredibly ornate mansions. It was speculated that had Mackay built on the windy hilltop, Louise might never have thought to leave. Perhaps she would have remained if he had built next to Flood's stately and impressive brownstone, encircled by a thirty-thousand-dollar brass fence that required an employee to polish it all hours of the day to keep it glistening. However that may have been, after three years in San Francisco, she was not satisfied. Her eighteen-year-old sister, Ada, was ready to return to the United States from Europe, and Louise had decided that daughter Eva, who was turning sixteen, and the boys, John Jr. and Clare, would benefit from living in a cultured, aristocratic society. Louise wanted to be part of a world more grand and eloquent than was to be found in the West. Her family had taken her from New York when she was quite young, and Theresa Fair noted that Louise's heart seemed to be set on returning. At the beginning of June 1876, Mackay and family left for the East.[19]

The economic elite was socially and politically dominant in New York in the second half of the nineteenth century as in no other place or time in American history. The Vanderbilts, Astors, Rockefellers, and Morgans still symbolize the era. They and their ilk owned capital without working manually, shared culture and identity, and were extremely class conscious. Their self-anointed uniqueness was institutionalized in clubs, associations, and debutante balls.[20] John Mackay was sought out by some of that class when the family traveled to New York. Despite that and the fact that his private secretary, R. V. Dey, who came from one of New York's old families, wrote introductory letters, Louise Mackay could not break through the barriers erected by the privileged class. She bought her way into a special edition of the book *The Queens of American Society* to no avail.[21]

The Vanderbilts' patriarch had begun his career as a ferryboat master. The Astors' elders had engaged in acquiring beaver pelts. Others got their starts as Yankee traders or linen salesmen. Yet by the 1870s, the status of their society could not be maintained if the wife of every successful capitalist to come along was allowed access. Ward McAllister later identified New York society's smart set as the "Four Hundred"—the number of guests that would fit in Mrs. William Astor's ballroom. The group encompassed 213 families, old New York names who could trace their lineage back at least three generations. They commanded the social scene in the city as well as the Newport, Rhode Island, community of million-dollar cottages in the summer. One gentleman's definition of American aristocracy was "inherited wealth." To be embraced by the inner circle, also known as "the world," one needed money, brains, tact, and time—all of which Louise possessed except for the last. It was best to let fashionable people seek out a newcomer rather than seem to be rushing too quickly to claim a place. McAllister noted: "Nothing is to be gained by trying roughly to elbow yourself into society . . . for when such a one has reached it, he will find its atmosphere uncongenial and be only too glad to escape it."[22]

Louise's endeavors to become part of the New York elite were made more difficult because the country was mired in the depression that began in 1873. In the economic crisis of 1857, the influential merchants and manufacturers had offered the poor relief and jobs. In the 1870s, instead of assistance, upper-class New Yorkers sought new ways to protect their property. They built boarding schools and museums, redecorated their houses, and held fancy balls. The depression strengthened their collective identity, creating a wider separation between the classes.[23]

Louise's actions in attempting to join New York society were awkward and, in the end, humiliating. Mackay wrote to three business acquaintances in New York. Each responded in the same fashion. They were glad to hear of his wife's arrival and stay at Everett House. If there was any service they might render, let them know. Each mentioned further that his wife was out of the city at present but would certainly call at Everett House upon her return. Silence followed the letters.

A woman whose husband was interested in talking with John about mining investments was coaxed by the husband into inviting Louise to an afternoon social at her home on fashionable and exclusive Washington Square. When Louise arrived, she was told that the hostess was out, although there were carriages with drivers lining the street, apparently waiting for their mistresses. As Louise's hired carriage circled the park to return to Everett House, she watched two ladies arrive, step out of their carriage, and enter the residence from which she had been turned away.[24]

Mrs. Paran Stephens, an acquaintance from San Francisco, offered to host Louise and introduce her to society if John would advise Mrs. Stephens regarding the San Francisco Stock Market. Louise agreed, and John reluctantly acquiesced. But at the coming-out party at Mrs. Stephens's, Louise overheard socialites commenting about her earlier failed visit. "Turned away at the door of course," said the first. "What else can one do if we're not to be drowned in a sea of gold—or silver, I suppose I should say." Another added, "You might say brass."[25]

Acting against Louise, besides her lack of accepted lineage, was her Catholicism, seen as the religion of the serving class, and the impossible circumstance of her personal history—seamstress in a California mining camp. Her husband's Irish heritage, and the fact of his having succeeded through manual labor, provided other reasons for the privileged class to shun her. Even at the turn of the century, after Clarence Mackay married socialite Katherine Duer and Virginia Fair married William K. Vanderbilt, the New York social critic Van Gryse, while acknowledging San Francisco as America's least provincial city after New York, added: "Californians don't carry bowie knives and derringers round in their boot now, but there is a sort of aroma of it clinging to them." It was too much for Louise. But instead of retreating, she made a move that revealed a social daring and tenacity equal to her husband's enterprise. She abandoned New York for an even more prestigious stage—Paris.[26]

As for Mackay, he served as a mines and metallurgy judge for the Centennial celebration in Philadelphia. Among the 2,129 metallurgy exhibitors, the Firm's display was an award winner in the "silver ores, gold bearing" category.[27]

As soon as his duties were completed, Mackay accompanied Louise and the children to Paris. There he purchased the "hotel," as mansions in France were called, at No. 9 Rue Tilsit. The two-year-old house fronted the Arc de Triomphe. House and grounds occupied a square in the exclusive neighborhood. It was four stories in Modern Renaissance style, with a mansard roof. There were two massive gardens, and the entry was a glass-enclosed court. When the arrangement for the purchase was completed, Louise's parents and sister came to live in their own suite of rooms.

Mackay, further aiding his wife's social aspirations in a demonstration of display that he steadfastly avoided for himself, gave Louise a necklace of matched sapphires set in diamonds. The pendant sapphire was the size of a pigeon's egg. It was the beginning of a sapphire collection that he intended to be the finest in the world. Then Mackay sailed again for America.[28]

His return to the West Coast was trying. At the same time the Firm was being attacked by "bears" from outside, an attorney and stockholder, Squire P. Dewey, assaulted them from inside. And in Mackay's absence, Fair had assailed him personally.

Fair made a statement, published on August 1 in *The Stock Exchange,* that Mackay had plundered the mine. It was Fair's as well as the Firm's policy to leave reserves of ore on all levels as tunnels were opened. Fair wrote: "It has been my constant purpose to so unity [*sic*] the higher and lower grades of ore as to make the average yield conform to an average standard value." He castigated his partner while ignoring the circumstances under which he had worked. "The mine has been managed in such a way, which was injudicious to say the least," stated Fair. "The richest ore was being extracted and run through the mills, the poorer quality left in the mine. Fancy runs were made and an enormous amount of bullion turned out, which of course could not be kept up without detriment to the mine." The correspondent commented: "Col. Fair resolved to correct this at once, and did so."[29]

But there was something more in Fair's chastisement of Mackay. James Keene's bear attacks centered on the weakness of ore on the lower levels of the Con. Virginia. He alluded to the pinching off of veins and the impoverishment of their quality below 1,550 feet. These observations were diametrically opposed to the claims made by Fair in the previous Con. Virginia yearly report. Fair now faced the reality that the ore at the lower levels would be practically mined out by the end of the year. When Keene and an investor from the Comstock's early days, George D. Roberts, came to inspect the mine at the end of July, Fair blamed Mackay in order to turn criticism away from himself. The Comstock citizenry was amazed that Fair would publicly denigrate his partner, but Mackay's reaction was typical: he did not respond.[30]

Fair's assertions reinforced the market bears' claims and gave impetus to more of the same. On July 17, 1876, the *San Francisco Chronicle* assailed the Firm. The *Chronicle* of the early 1870s was something of a reformist paper, having published several editorials (probably authored by the progressive Henry George) attacking California land monopolies. In the mid-1870s it turned its attention to the dangers of stock manipulations. But the attack on the Firm was galling, since previously it had differentiated between the Bank Ring's management, which it termed dishonorable, and the Firm's practices. Eighteen months earlier, the paper had lauded the Firm for advising poor friends to invest rather than engaging in frauds and hording the profits, as had been the rule with other management groups. It commented that the Firm had brought about new enterprise and assured "that San Francisco is not to be longer at the mercy of the old junta of mining sharps and money-grabbers." Supporters believed the *Chronicle's* new criticism was merely a continuation of its reform policies; detractors attributed it to the paper management's having lost money in Con. Virginia stocks.[31]

The *Chronicle's* article had been signed "Unknown," but it was widely believed that San Francisco investor Squire P. Dewey had written it. The criticism may have been the last straw for Flood, who quit attempting to shore up the market. Bears continued to sell and collect profits as the market fell. The decline went on even though the California mine came into production on April 1, began paying dividends shortly after, and earned $12,505,320 by the end of the year.

Keene, having amply padded his fortune in the bear market and arranging to move to the East Coast, visited the mines again at the beginning of August. This time he told the papers he believed prices would improve and now thought they were a good investment. But he also warned that markets were always unstable, and stocks fell further. At the end of 1876, claiming the shaft needed repairs, the Con. Virginia failed to pay a dividend. It was the first time it had not disbursed monthly payments since commencing them in May 1874. Con. Virginia stocks had split in March and sold for $90 a share. In December they plummeted to $35½. California mine shares fell from $95 in March to $43 in December, and Ophir stocks plunged from $76 to $20.[32]

In 1877 the Firm controlled much of the Comstock Lode. At the end of July that year, 3,156 men were employed in Comstock mines: the Firm controlled the Con. Virginia, California, Ophir, Mexican, Best & Belcher, Gould & Curry, Savage, Hale & Norcross, Yellow Jacket, and Utah mines, utilizing a total of 1,919 men; William Sharon's interests—six mines—employed 534 men. Of the four other significant employers, only one employed more than 100 workers.[33]

Although S. P. Dewey denied authoring the newspaper article attacking the Firm, there was reason to believe he was responsible. Dewey was involved in real estate and mining investments. A prominent stockbroker remarked that Dewey was "somewhat noted among his friends and acquaintances for his self-importance and rather pompous manner." In January 1877 he began a frontal assault on the Firm's management practices. After the fire that destroyed Virginia City and before the extent of the damage to the mines was known, Dewey had approached Flood to ask if the Con. Virginia's next dividend would be paid. Flood told him he was uncertain, referring him to Charles H. Fish, the mine secretary, who at least could tell him the amount of money on hand. Fish told Dewey the amount of cash but neglected to include the amount of bullion. Dewey, assuming the cash was not enough to pay the next dividend, hurriedly sold his stocks at a substantial loss. When the bullion was converted and the regular dividend paid, stock prices rose. Dewey blamed Flood for misleading him and causing his loss.[34]

Dewey, who claimed to have lost $52,600, wrote to a friend of the Firm asking him to make a proposal to Flood and O'Brien that would allow Dewey to reclaim his loss. "I claim that that loss ought in justice to be made good to me, hence my intention to pursue it. . . . I will sell to your friends 1,000 shares of California which I hold, for the above named sum of $52,600, which will divide the loss, with the understanding that thereupon our old social relations shall be renewed . . . and that counsel or friends—on either side, however intimate—shall know nothing more of this matter." Papers supporting the Firm later labeled the offer blackmail. Flood rejected the settlement, and the battle was joined.[35]

At the Con. Virginia annual stockholders' meeting on January 11, 1877, Dewey commented on the secretiveness of the mine's managers. If the facts regarding the mine had been released a month earlier, he argued, shares would not have plunged as they did. Flood rose and said: "Mr. Dewey, you have been my enemy for over a year, and you are nothing but a stock gambler." Dewey responded: "I have lost $50,000 on a falsehood told to me by you."

Mackay interjected that he had not kept information from the public, he did not believe Flood had, and Fair's weekly reports were open to all. Dewey and Flood again engaged in a heated discussion.

"Come, come, gentlemen," said Mackay, "If Mr. Dewey will gamble in stocks he must expect to lose once in a while, and at any rate this is no time to bring up personal matters. I want to say now—and not only for the benefit of Mr. Dewey, but all the gentlemen here present—I am the largest individual stockholder in the company, and have more at stake than any man in it. I work the mine with Mr. Fair in Virginia, and it is simply impossible for us to tell two or three days or a week ahead what the mine may develop." He explained that hard rock, cave-ins, bad air, and other variables constantly altered calculations.

James White, an investor from England, came forward to say that he represented others in Britain. They were dissatisfied, White said, because affairs of the mine were not disclosed openly. Mackay now became upset, replying that an Englishman could know nothing about the working of an American mine and that "those who were so much given to blowing about the mine ought to come to Virginia and try to run it themselves." He turned to Dewey: "Now, Mr. Dewey, my dear sir, if you will come up to Virginia you shall have all the information you can possibly get."

Dewey said he and the other stockholders wanted the management to do something to give them more confidence. When Mackay listed expenses— mill costs, wages at $4.50 to $10 a day, expenses in handling the ore—Dewey said everyone knew all that. Mackay again detailed obstacles in the mine such as water and ventilation, commenting: "Then if there is any delay people say there is something wrong, and get mad."

The vote for trustees was held, and the Flood and O'Brien ticket won 483,136 shares to 32,335 for the opposition.

Dewey then read a resolution, which offered three proposals. The first was that the directors be required to publish monthly statements of receipts and expenses. The second was that the trustees give the secretary copies of all surveys, plans, and drawings of the underground workings and keep the secretary apprised of the condition and profit of the mine. The first proposal was rejected, the second accepted. During the lengthy discussion, Mackay commented: "It is very hard to reply under the circumstances. . . . I defy any gentleman to say that there has been any dishonesty in the management of the Con. Virginia mine." No one challenged him.

Dewey's last proposal was that the trustees purchase or build mills to be run by the mining company instead of using those owned by Flood, O'Brien, Mackay, and Fair. Dewey had spoken to someone who believed new mills would pay for themselves every ninety days. Mackay responded that he was misinformed: "No mill can be cleared short of a year or 18 months. Then we must consider the risks that are to be run of the property being destroyed. You cannot build a mill that will crush 300 tons of ore for less than $450,000 or $500,000." Several stockholders spoke, all rejected the idea of the mining company's trying to run mills, and—despite lengthy arguments by Dewey that the Firm was appropriating millions for themselves without stockholders' consent—the motion was voted down.

Dewey brought up mine tailings, the residue separated from the ore in milling, arguing that the Firm could profit from them. Mackay said they would sell the tailings if anyone would want them. "These tailings ran to waste in the Carson River for 10 or 12 years, until Mr. Hobart and myself conceived the plan of running them down to Big Flat and saving them. The plan cost $130,000. My object in taking them down was, as I have a very large family, to give them to the little fellows to work when I am gone."

The stockholders almost unanimously supported Mackay. In response to Dewey's charges and inferences, they offered two new resolutions: one thanked the trustees for their management during the past year; the other ratified all that the trustees had done. The discussion again became heated. Dewey called the resolution thanking the trustees "a whitewashing resolution." White, at one point, called out "Monstrous! Monstrous! Monstrous!" When he tried to remark further, Mackay blurted out that to suggest dishonesty ill became White as an Englishman and that during Mackay's last visit to London, there was a scandal that made it difficult to find an honest man there. White objected. Someone from the crowd said he thought Mackay ought to show some courtesy, and the talk quieted. Each of the resolutions supporting the Firm passed, and the meeting was adjourned.[36]

The display of Mackay's temper, once described as quick but quickly over, was predictable. What was surprising was his expression of contempt for the English while responding to White. In an era of significant identification with national origins, widespread discrimination against the Irish, and persistent English-Irish contentiousness, Mackay seemed consistently impervious to race. His charities and his hiring and treatment of workers were evenhanded. He later hired an Englishman as general manager of his cable business, and his second home would be a great estate in London. Years afterward, when he was anonymously accused of prejudice against the English and against Jews, the *New York Times* pointed out that he regularly dealt with English associates in his role as a director of the Canadian Pacific Railroad and that when he sold his western bank to a Jewish syndicate, he retained a large interest and served on its board, frequently bringing him together with Jews.[37]

The day after the contentious meeting, the *San Francisco Chronicle* and several papers with smaller subscriber markets came to markedly different conclusions than the stockholders who had voted to support management. The *Chronicle*'s article, entitled "Rights of Stockholders," supported all three of Dewey's propositions. Three other papers took the gloves off to assail Mackay. The *San Francisco Post* commented on his "hotheadedness": "He brought to his aid bad grammar, ungentlemanly retort, and all the conceit which the consciousness of great wealth can inspire." The *San Francisco Mail*, in assailing him, featured an air of crass condescendence. It concluded: "[If Mackay] is really so tired of 'me and Fair's' ineffectual struggles to give satisfaction, let him resign, and permit the stockholders to try some one else for a little time." The *San Francisco News Letter* accused him of being a "bulldozer," saying his treatment of White "might have been expected from a scrub, who raised from nothing, had attained to nothing except dirty purse pride." (Interestingly, the *News Letter* a short time later published an article that featured a caricature of Dewey with a devil's tail, assailing him as a blackmailer.)[38]

These attacks were met by a defense of Mackay and the Firm by another of San Francisco's leading papers, the *Alta,* and later by the daily mining journals the *Stock Report* and the *Stock Exchange.* On January 13 the *Alta* commented that a portion of the press was attempting to denounce the West Coast's enterprises and vilify its most able men. It continued: "The attacks on the bonanza mines and the venomous falsehoods with which their officers have been pursued, afford a singular instance of depravity on the part of certain newspapers." The *Alta* pointed out that the Con. Virginia had raised over $37,000,000 in bullion, of which $28,079,080 had been paid in dividends, and between the two bonanza mines, of the roughly $50,000,000 produced, nearly $38,000,000 was dividends. This was achieved, it continued, while explorations were pushed, a new shaft was built, new mills were erected, flumes to supply timber were constructed, and all other necessary operations were carried out.

On January 14 the *Alta* refuted the attacks on Mackay: "We believe that no other of the great millionaires of our times is so noted as Mr. Mackay for conduct so inconsistent with purse pride, for simplicity of manners, cordiality of relations with old friends, no matter how poor, and strict attention to hard labor."

Virginia City's *Territorial Enterprise* also weighed in on the 14th, commenting that while the press of San Francisco was accusing mining companies of rascality, there had been no change in mining and milling operations in years, a fact all investors would have known before buying stocks. The *Enterprise,* owned by Sharon, Mackay, and others, then delineated how Sharon had formed the Union Mill and Mining Company, thereby saving mining on the Comstock. The paper did not distinguish between the Bonanza Firm, which paid large percentages of its proceeds to stockholders, and Sharon's company, which managed its mines primarily so the market might be manipulated. The *Enterprise* article quoted Mackay, restating the challenge he had issued in the meeting: "Why don't some of you people come up [to the Comstock], take off your coats and see what sort of a showing you will make."

With all the commotion, stock prices rose in January 1877. Con. Virginia sold for $55, California for $53, and Ophir for $56. Unfortunately for the market, the big bonanza was playing out. Values sagged thereafter and, with the exception of Ophir's brief rise to $98 in September of 1878, none of the mines reached $50 again.[39]

The *Enterprise,* the *Alta,* and the mining journals continued their defense of the Bonanza Firm into the spring. In April 1877 the *San Francisco Stock Report* called the *Chronicle*'s editorial stance an assault on the market, the people, and the entire state. The *Stock Report* accused the paper of using "the same misrepresentation of facts" espoused since "the *Chronicle* was sold out on its margin," saying that the *Chronicle*'s comments were intended to victimize the people, and concluded that the people at last could see through "the *Chronicle*'s scheme of robbery." At the end of the month the Englishman James White, at Mackay's invitation, traveled to Virginia City and inspected the mines. He brought his own expert and after the visit pronounced himself fully satisfied. The *Stock Report* commented: "So explodes another *Chronicle* canard."[40]

CHAPTER 9

Status and Scandal

LOUISE MACKAY was the first of the ultrawealthy Americans to conquer European society. For over a quarter of a century she lived as an émigré, hosting magnificent fetes attended by those of rank in the European social order. Ostensibly she began living in Paris so her children would benefit from a superior education, but certainly Louise's motivation for the move included her desire to gain a place among the social elite. In the 1870s New York society reporters began heralding festivities of the city's elite. It became acceptable to flaunt one's wealth. Debutante balls, flamboyant dinners, and other ostentatious public displays followed. The fact that the American press, including the New York dailies, trumpeted Louise's European triumphs must have been sweet vindication for her after her ignominious stay in the Empire State. Coverage of her activities insured that imitators would follow. Few succeeded so well as she.[1]

The four-story Mackay mansion at 9 Rue Tilsit was decorated in the latest chic. The era's parlors were essential to one's status, displaying the owner's learning and sense of art, culture, and design. Louise would not be outdone in the creation of a home-fashion statement. The suite of small rooms at the entry was furnished Pompeian style, with frescoes on the walls and a mosaic floor. There was a small salon, hung in pale blue satin, where she might receive distinguished guests. A ballroom was decorated in white and gold. The dining room was dark woodwork, relieved with golden lines and gold brocaded curtains. The many corridors were richly carpeted, said to give an air of snugness rather than wealth. The second floor was equal parts living quarters and aviary. In the vestibule a large parrot spoke to himself with a vocabulary that included English, Spanish, and French, while the cages of various species of songbirds filled other rooms.[2]

Upon moving into the house, Mrs. Mackay began to issue invitations to those who might help her achieve social advancement. As is generally the case in situations where social status is involved, there was a group in the American colony in Paris who opposed her. As in New York, she was despised by some of the old guard because she was a member of the nouveau riche, whose wealth eclipsed their own. But the *Territorial Enterprise* reported that she won over others by her benevolence. "There is not in all Paris," it said, "a more charitable woman than Mrs. Mackay."[3]

John Mackay's economic status and his support of the dominant Republican Party caused politicians, including American presidents, to seek him out. He served as a member of the Republican National Committee, hosted ex-president Ulysses S. Grant in 1879 and President Rutherford B. Hayes in 1880 when each visited Nevada, was later appointed a special ambassador to the coronation of Russia's Alexander III by President Chester Arthur, and even visited Presidents Grover Cleveland and William McKinley at the White House. Because of her husband's affiliations, upon Louise's arrival in France she was introduced to Mrs. Edward Follansbee Noyes, the wife of the American minister. Mrs. Noyes introduced her into the American society there. She also met and gained acceptance from Isabella, the exiled queen of Spain. Ellin Berlin, Louise's granddaughter, wrote that while both acquaintances were welcoming and friendly toward her, Louise kept them at arm's length; intimate society was reserved for her mother, sister, and almost grown daughter. Berlin commented: "She did not require love or friendship from other women. Friendly affection and easy kindness were all she would ever ask or need from her new acquaintances."[4]

In November 1877, a particularly rainy and foggy season, ex-president

Grant and his wife came to Paris. Entertainments had been arranged for every night of their stay, and Louise had secured the last night, the 21st, to host a dinner and grand ball honoring them. It was because of John that the affair was scheduled, and he could have given her no finer gift. If the evening were a success, her social position in Paris would be assured.[5]

Mrs. Mackay suggested to the municipal authorities that the Arc de Triomphe be decorated in honor of the general's visit. When her request was denied, she remarked in jest that she might consider buying "their old arch." The Paris newspapers were publishing daily stories about the Americans, and when this item appeared, it was reprinted across Europe and America. The quip, interpreted as either feisty or arrogant, awarded the budding socialite a large measure of celebrity on both sides of the Atlantic.[6]

On the evening of the event the facade on the Rue de Tilsit was illuminated, lighting the street for arriving guests. The glass courtyard was filled with plants and flowers from the south of France. Her birds, in an aviary, and bright fish, in an aquarium, were the main attraction in the Pompeian suite. The twenty-four menus for the Grants and the American legation and consulate, who made up the dinner guest list, were engraved on small silver tablets. At the reception and ball there were over three hundred guests, including the Marquis de Lafayette and French president MacMahon. Although the guest of honor retired early, the dancing continued until 4 AM. In a firsthand account William Ralston Balch, an early Grant biographer, wrote that for the American colony and fashionable society "it was the greatest sensational event of the season."

In the United States during the period, there was a fascination with European society, and henceforward Mrs. Mackay's social triumphs were certain to earn prominent mention in social columns across America. Her life was also idealized in fiction. In *Under the Tricolor,* a novel by Lucy H. Hooper, dedicated to her, the Louise character was described as good, generous, and unostentatious, "though her husband's wealth surpasses all ordinary methods of computation."[7]

John Mackay had missed the celebration honoring Grant, and it was just as well. In another genteel romance Ludovic Hlévy, a popular French novelist, also patterned a main character after Louise Mackay. There was a description in the book of the husband, a silver miner from America "with a hundred thousand francs a day to spend." At an exquisite ball given by the heroine, her husband appears. "He bowed to the right and left at random. He did not enjoy himself, I assure you. He looked at us and seemed to be saying to himself: 'Who are all these people? What do they come to my house for?'" This description was taken from life. It was reported that Mackay avoided social functions whenever possible. When that was impossible he would leave early, retiring to the upstairs to play billiards.[8]

His reluctance to participate more fully at his wife's large gatherings does not seem to have been a reflection on their marriage relationship but an aversion to large groups and the European aristocracy. To Mrs. Mackay's mortification, he was at times disdainful of the nobility who frequented the house, alluding to them as "bums and parasites" who ought to go to work.

Mrs. Mackay's parents, Colonel and Mrs. Hungerford, lived with her for long periods of time. Colonel Hungerford used Mackay money to invest in various schemes and promotions without success. In an anecdote about him related by Oscar Lewis, Mrs. Hungerford returns home to find the cook missing. "She was drunk and I dismissed her," explained Hungerford. "You're drunk yourself," said the wife. "That may be, m'dear," he replied importantly, "but I won't have two people drunk in the house at the same time." Mackay was generous and gracious toward his in-laws, although Alexander O'Grady said that while careful not to let the ladies know, "he never took the Colonel seriously."[9]

Alexander's mother, Alice O'Grady, was the Mackay children's nurse before becoming manager of the entire house in Paris. Alexander was the same age as John Jr., the Mackays' oldest son. Alexander, later a member of the San Francisco Bar, slept in the same room with John Jr. They were best friends, sharing a tutor and clothes. Alexander had a brown pony like John's and was treated in every way like another Mackay child, to the point of being introduced by Louise as her adopted son.

On Mackay's frequent visits to Paris, he always brought presents for the children. He would play with them and joke with everyone. Once while playing croquet with the little boys, Mackay became upset with John Jr. He took his son over his knee when Alexander, to protect his friend, hit Mackay with his mallet. Mackay, with an eye swelling and blackening, dragged him to his mother, sputtering: "Alice, if you don't take this damned brat out for a drive, I'll kill him." They became great friends afterward. Alexander believed Mackay liked him more because of his spirit.

When Mackay went for his morning walk, he took the two boys with him because, as Alexander explained, "he liked our prattle." If he needed to converse with anyone in French, he had the boys translate, refusing to make an effort to learn French himself. The French chef's meals were said to be an abomination to Mackay, who could not wait to get plain American food—corn beef and cabbage, for example.[10]

Late in May 1878 Mackay visited the great Paris Exposition. He had tried to get Congress to appropriate money for a Pacific Coast mineral exhibit at the show. When they balked, he funded it himself. Over 1,000 specimens from California, Nevada, and the Pacific Territories had been collected, cataloged, and labeled, and those from the premier Comstock mines were said to be remarkably rich in gold and silver. But the mining exhibit was overshadowed by a display of half of a silver set created by Tiffany's of New York. It was the Mackay setting, a gift for Louise, which when completed contained 1,300 pieces made from a half ton of Con. Virginia silver. Two hundred silversmiths had worked two years to finish the elaborately designed set. The collection included ornate carving and serving utensils; sets of 48 heavy flatware pieces; items of occasional use such as nutpicks, ice-cream, fruit, and egg spoons in sets of 24 or 36; candelabras holding more than 500 ounces of silver; ornate bowls; silver sleighs for serving claret; and a tea set of silver said to have been carried out of the mine by Mackay himself.[11]

Striking at the exposition, the service did not impress a critic for the *New York Tribune*, who called it art "running riot." He believed it was intended to astound the beholder with its costliness and the skill of the makers. Noting that ornament for ornament's sake is bad, he commented, "There never was a more signal proof of the principle than this over-done service." This, like most of Mackay's great expenditures, was done for Louise. His life was lived in almost austere plainness, her life largely in display for society.[12]

When Mackay spent for himself, it was as a patron of the arts. He became known for his collection of paintings, on one European trip spending $150,000 for pictures in Rome and $50,000 in London. He bought *The Negro Barber* by Leon Bonnat in France for 22,500 francs and paid 51,000 francs for another famous French work, Gerardt Dow's *The Fish Merchant*. In 1883 the *New York Sun* reported that Mackay paid $500,000 for Sir Philip Miles's collection of Leigh Court art and said, "He has been in negotiation with Lord Lansdowne for a series of the Italian masters of the highest renown." But even most of these purchases remained in Europe in the mansions in which Louise resided.[13]

In the United States, Mackay split his time between the Comstock and

San Francisco. In the city, as in Virginia City and later New York, he frequented the theater. At that time the tragedian Edwin Adams had returned from the East. He had debuted in San Francisco, playing Hamlet in 1867, and performed at the California Theater in 1869 before gaining stardom by touring the East. The great Mercutio of his era, he was sick with tuberculosis in the late 1870s and penniless. San Francisco actor and theater manager John McCullough reported that when Mackay heard of Adams's plight, he sent the sickly performer a check for two thousand dollars. In the accompanying letter, Mackay said it was partial payment for the many occasions he had enjoyed the actor's performance. Adams' health improved. He began performing again and regained his former prestige. McCullough said that until his death the Mackay letter was Adams's most treasured possession.[14]

Mackay was said to have given the lionized Lawrence Barrett twenty-five thousand dollars and comedian and character actor Billy Florence fifteen thousand dollars, and he frequently helped McCullough. Although these three repaid their debts, Mackay thought nothing of giving when it was certain the debtor could not. When the young English matinee idol Henry Montague was stricken with consumption in San Francisco, Mackay secretly paid his bills. Montague died during a performance from a "hemorrhage of the lungs," and Mackay paid for the body to be shipped to the Montague family.[15]

One of Mackay's acquaintances became a friend who later would become a business partner. Edward S. Stokes was generally held to be unscrupulous, although a well-educated member of an old, aristocratic family. He was dashing and one of the handsomest men in New York when he arrived there at age twenty-four, one of the gilded youth. Stokes met Mackay upon coming to San Francisco to start life anew after his release from Sing Sing prison in October 1876. His crime, committed in January 1872, was one of the most spectacular transgressions of the era: he murdered Jay Gould's partner, "Jubilee" Jim Fisk. Although married, the rotund and ebullient Fisk had intimate relations with a host of ballet dancers, noting: "I'm the gander can take care of these geese." Fisk's favorite was Josephine "Josie" Mansfield. Fascinated by her brazen attractiveness and wit, Fisk made a public display of his attentions. He bought her extravagant jewelry and kept her in a four-story residence, with three servants, a carriage, and a coachman at her service. She kept the exquisite young Stokes, who was also married to someone else.[16]

At the end of November 1871, the scandal exploded in the public arena when hearings on a suit, *Mansfield v. Fisk,* began. The case charged Fisk with libel, and was brought because Fisk publicized the fact that Stokes and Mansfield had been blackmailing him. Josie had letters that alluded to connections between Fisk, Gould, Tammany Hall, and Erie Railroad villainies. In the trial the letters were suppressed as evidence, and instead testimony was elicited exposing the intimacy of Mansfield and Stokes. Stokes's proud family was humiliated.

When Fisk arranged that charges of extortion be brought against Stokes and Josie, Stokes could take no more. He ambushed Fisk in New York's Grand Central Hotel, fatally shooting the large man in the stomach as he came upstairs. In his first trial Stokes was sentenced to death and, for a time, was kept in the condemned cell of the Tombs in New York City. Stokes appealed, claiming that Fisk had been armed and looking for him, and only through evidence tampering and the perjury of witnesses had the "Erie Ring" obtained his conviction. After two more trials, he was finally jailed for manslaughter, released after three years, and restored to citizenship by President Arthur on December 1, 1884. When he came out of prison his appearance was still striking, although his hair had turned white and he seemed always on guard.[17]

Friends never understood Mackay's relationship with Stokes. One account of how the two came together involved the Victorine mine near Austin, Nevada. The Firm invested in it in the fall of 1880—purchasing 160,000 shares from Stokes. It was a failure, causing the Firm considerable trouble and expense. In March 1881 they deeded the mine back to Stokes and, according to Grant Smith, the transactions led to the friendship. A longtime Mackay associate told another story, saying that when Stokes arrived in San Francisco, he bought a fast pacer that Mackay admired. When Mackay inquired into buying the horse, Stokes gave it to him without charge. In return Mackay advised Stokes in mining, allowing him to make a fortune. In either case Mackay later financed Stokes's buying the successful Hoffman House Hotel and Restaurant, and they worked together in developing the Mackay telegraph system. They remained friends until an acrimonious lawsuit ended their association some fifteen years later.

Mackay had become cultured. He read a great deal, traveled often on business or to visit his family, and associated with educated men. He took at least one book with him on trips, was fond of biography, and always carried a volume of Shakespeare's plays. No matter the city, he attended dramatic productions and opera. But most often he was alone. For years, when in New York, Mackay reserved a round table for five in a corner of the Hoffman House Restaurant in case friends happened in. He always preferred to eat with others and was happiest when he had guests from the West with whom to dine. The meals were simple, and the others did most of the talking. Even with friends he was reticent, rarely engaging in conversation except to encourage them to talk. Early in his career, when fame began to crowd out his old associates, he complained more than once: "Why, even old Jack O'Brien won't come to see me any more!" Years later he commented: "I have had enough lonesomeness in the past to last me the rest of my life, and I never want to again eat or drink alone; and I want all my friends to understand there is always a place for them with me whenever I am at my own board."[18]

C. C. Goodwin said: "In natural bearing [Mackay] was imperious as a Caesar, with the walk of a trained soldier, but it was only in his bearing. As he mingled with men, there was not one look or word or gesture that was not winsome, unless some base nature crossed him." Eliot Lord called Mackay "a sober-faced man habitually." He commented that although Mackay had a good sense of humor and relished a good story or comical scene, he never laughed loudly. But he went further, saying that Mackay was "at all times and in every company, one of the most thoroughly natural men whom I ever knew."

Mackay never seemed completely comfortable with his wealth. It was an era when tremendous discrepancies were apparent, because cities had millionaires' estates at one end of the street and tenements at the other. The leading social reformer of the day was Henry George. In his philosophy wealthy individuals could build pyramids or burn their money if they wished, as long as they were not robbing or keeping others from their opportunity. Mackay would have agreed, but at times sentiment got the best of him. E. C. Bradley, a vice president with Postal Telegraph Company, told of an evening at dinner with friends when Mackay suddenly announced: "We ought to be ashamed of ourselves! Here we are living on the fat of the land, while thousands are starving." The next day he gave one of the friends a check for two hundred thousand dollars to distribute to the needy in the friend's name. Mackay never referred to the incident again.[19]

Mackay's partner James Flood was also generous. He advised friends he felt were adept at handling their own investments and bought stock shares due to rise for those less capable. He was known for his liberal contributions to charities, including a custom of sending checks to old peoples' homes and orphanages at Christmas, and for a gift of twenty-five thousand dollars to the suffering people of Ireland.[20]

In Flood's office in the Nevada Bank of San Francisco was a watercolor, *Changing the Shift*, that had cost the impressive sum of five hundred dollars. It featured Mackay, carrying his lunch bucket, emerging from a mine. From that office Flood tended to the finances of the Bonanza Firm's business. A wide assortment of people came to see him, asking stock advice, requesting loans, and proposing schemes in which he might invest. His clerk recalled that there were callers of every station, from General Grant to representatives of the U.S. Mint, women of title, bankers from around the world, and various cranks and pests.[21]

Flood had been a partner with his best friend, William O'Brien, since 1856. Although part of the private bank office was O'Brien's (he was on the bank's board of directors) and he had spent twenty-five thousand dollars for a seat on the stock exchange, O'Brien rarely involved himself in either business. As the silent partner of the Bonanza Firm, he explained his success by saying that he had caught the tail of a kite and hung on. O'Brien's most valued possession was an engraved silver trumpet given him in the mid-1850s after completing his term as foreman of a volunteer fire brigade. Although most pundits have disparaged O'Brien's contributions to advance the Firm's business, Oscar Lewis noted that he played a valuable role. Respected by the others, he served as a mediating force when their sometimes volatile personalities clashed. He also assuaged the feelings of the community toward his partners. Lewis commented: "O'Brien's geniality and unassuming manner did much to soften the rancor of a suspicious and hostile public."[22]

Flood was devoted to O'Brien, saying there was nothing he had that Billy could not share. Although frequently caricatured as an amiable drunk, the bachelor O'Brien had a distinguished demeanor. Novelist and California historian Gertrude Atherton said that while the Floods were plain people, O'Brien's forebears must have been well bred. She described O'Brien as entering a fashionable ball with one of his sisters and her daughters, whose stately grace and grand air caused the local socialites to "gasp [with] admiration." O'Brien provided homes, luxuries, and travel for his sisters and their families. One of his sisters' daughters married a viscount, another became a social leader in Washington, D.C., and his other sister's grandson became a Pulitzer Prize–winning author.[23]

As for O'Brien himself, he spurned the grand barrooms of the city's palatial hotels, the Lick House and the Palace, where other nouveau riche passed
their time. He preferred the back room of McGovern's Saloon on Kearny
Street. There he spent afternoons taking or losing tricks playing pedro with
his friends. He kept a stack of silver dollars on the table, and any of his acquaintances in need was free to take one. When the stack ran out, he cashed
another twenty-dollar gold piece from his bottomless pocket.

O'Brien's constitution was not strong, and in the winter of 1877–78 his
health deteriorated to such an extent that he commissioned a mausoleum
to be designed for him. He was fifty-three years old. In the spring he moved
north to San Rafael, hoping a climate milder than San Francisco's would
restore him, but he continued to fail, and on May 2, with family and James
Flood at his bedside, he died. He left $50,000 to orphan asylums. The bal-
ance of his $10 million estate went to his sisters and their families. His gen-
erous nature was evinced by a list of $280,000 in unpaid promissory notes
from those to whom he had lent money. He was a few years older than his
partners (six years older than Mackay) but lived as if he were younger. It
was said he would have stayed young if he had lived to be a hundred. The
normally reticent Flood commented: "Billy was my partner once. He is my
partner now, will be my partner forever."[24]

As with many financial success stories, the Bonanza Firm's achievements
were accompanied by litigation. The Firm was involved in two significant
cases in the late 1870s, and Mackay personally was ensnared in a third.
The increased Nevada mine tax meant to fund Sharon's railroad, fought by
the Firm since February 1875, had been a prime topic in the newspapers,
with Sharon's *Territorial Enterprise* defending the law: "It was through its
operations that the state Treasury was put upon the firm and prosperous
basis that it has enjoyed ever since the passage of the measure." As to the
charge that Sharon had instigated the change, the paper denied it to a point:
"Mr. Sharon was not instrumental in the passage of the law of 1875; but if
he was, he performed a meritorious act that proved the financial salvation
of the state." The tax was also disputed time and again in the Nevada legislature. Compromises, including assessment reductions, payment without
the amount of accrued interest, or payment to be used for schools and the
general fund—but not for the railroad—had been defeated in the state house
and state senate or, when passed, vetoed by the governor. After all potential

legislation was defeated, the Firm lost in the courts as well. In the end they paid the tax in full, averaging $250,000 per year.[25]

The Firm was bested in a suit brought by Squire Dewey as well. In 1882 Robert Louis Stevenson commented that the stock exchange was, in a peculiar manner, the heart of San Francisco, "a great pump we might call it, continually pumping up the savings of the lower quarters into the pockets of the millionaires upon the hill."[26] Dewey was determined to siphon some of that money back down the hill. He had a montage of complaints against the Firm that were finally condensed into three great claims totaling nearly $40,000,000. The first two suits, which were eventually dropped, involved the Firm's timber profits and income from the mills that Dewey thought should be shared with stockholders.

In testimony given in the U.S. Circuit Court, Flood acknowledged that he, Mackay, Fair, and O'Brien were the principal stockholders of the Pacific Wood, Lumber and Flume Company; the Pacific Mill and Mining Company; and the Virginia and Gold Hill Water Company—the exclusive sources of wood, milling, and water requirements for the Bonanza mines. When asked, "Is there any other corporation from which the company draws supplies of any character of which Mackay, Fair, Flood and O'Brien are not the trustees and principal owners?" he replied, "I don't know of any." Many millions of dollars came to the four partners by way of the arrangement, and Dewey asserted that much of it was in the nature of skimming and fraudulent insider transactions.[27]

In speaking of the operation of Comstock mills, Eliot Lord pointed out that awarding contracts to the lowest responsible bidder was rarely done. Instead competitors utilized secret underbidding, gifts to mine officials, or kickbacks in the form of percentages of milling costs.[28] It can be assumed that would be the case in supplying mines with timber and cordwood as well. Because the charges levied by the Firm's subsidiaries were the same as the operating expenses paid by other mines, pundits have never called their lucrative profits exorbitant. In fact, Grant Smith said: "The members of the Firm played a fairer game than any other group in control of Comstock mines." By June 30, 1880, Con. Virginia had paid $42,390,000 in dividends while assessing stockholders only $411,200, and the California paid $31,320,000 without assessments. The profits of each of these more than doubled those of any other mine, illustrating the management of the Bonanza mines. An example of dubious management working for other than stockholders' interest was that of the Yellow Jacket mine, controlled during its greatest production by Sharon. It worked in good ore many years and paid $2,184,000 in dividends but assessed its stockholders $4,638,000.[29]

Dewey's third claim, in which he would win a settlement—far below the amount of the filing—came to trial in December 1880 and was decided at the end of March 1881. Suit was filed for $10,429,068 in the name of John H. Burke, accusing Flood of acquiring certain shares of the Con. Virginia mine in a manner that defrauded the stockholders. Early in the development of the mines on the Comstock, litigation was interminable. Mackay, Fair, Flood, and O'Brien knew to avoid such confrontations by clearing titles before digging. By all accounts they had done a comprehensive job before undertaking the Con. Virginia enterprise. But the plaintiff now contended that when the Firm's attorney, Solomon Heydenfeldt, bought the Kinney claim (which accounted for 50 feet of the Con. Virginia), 12½ of its feet was purchased for the benefit of Flood and the Firm. The resulting profits, the suit claimed, should not belong exclusively to them but to all who held the mine's stocks. Other stockholders were solicited by Dewey to take part in the litigation, but none did. The *San Francisco Bulletin,* on May 25, 1878, suggested that the case was akin to blackmail. It commented: "The firm of Flood & O'Brien, or any other prominent firm that may be assailed, will confer a lasting benefit on the community by fighting it to the bitter end."

When Plaintiff Burke was called to the stand, it was pointed out that the purchase of the Kinney ground had occurred in 1872 or 1873 and that Burke had bought his one hundred shares of stock in the spring of 1878. He had learned of the Kinney purchase from his attorney and refused to answer regarding where he had obtained the money to buy his shares, leading to the assumption that it had been provided so his name could be used in the suit. Although the court found that the transaction in question was not "actually or willfully fraudulent," it was determined that $930,000 in earnings should, indeed, have been divided among all stockholders.

Rather than face the inevitable appeals on the horizon, the litigants compromised. While the terms of payment were kept private, since no other stockholders had joined in the complaint, they were certainly far less than the total. Burke complained about the settlement, saying his attorneys and Dewey had talked him into accepting it "much against my desire."[30]

On December 27, 1878, a more startling complaint against Mackay was brought to public attention: John Mackay was named the defendant in a lawsuit, claiming damages of $200,000 and costs of the action, for the seduction of another man's wife. Mackay's fame, along with the unusual magnitude of the damages, combined to give the scandal two full columns of space in the *Chronicle*. William H. M. Smallman, the litigant, had married Amelia Hodgden Fritz Smallman the previous February after they had lived together for six months. They resided first in the Grand Hotel and then in the luxurious Palace Hotel.

The complaint alleged that Mackay contrived to alienate and destroy Mrs. Smallman's affection for her husband. It charged that between May 15, 1878, and October 23, 1878, the two had carried on an unlawful intimacy, stating further that the acts had ultimately caused the wife of the plaintiff to succumb to a mental illness. The complaint reported that Mrs. Smallman was in St. Mary's Hospital "in a hopeless state of insanity."

Mrs. Smallman was in her twenties. Her exact age was subject to conjecture, since she had reported it as nineteen for her first marriage in 1870 and twenty-four for her second in 1878. Her eventful personal history first signaled that something in the suit did not ring true. Her past included the dissolution of her first marriage to Henry Fritz, familiarly known as "Maguire's Fat Boy"—said to be a man of elegant leisure. A "persistent tendency to flirtation" on the part of the wife was a major factor in concluding the relationship. Afterward she fomented talk by traveling East in a palace car with a man named Livingstone, and subsequently she sojourned with a Parisian banker named M. Sellier. While unmarried she had had several live-in lovers, and once while she was occupying an apartment alone, her landlady had presented Amelia's boyfriend with a disheartening report of frequent male visitors at a variety of "unseasonable hours." Others of Amelia's amours, as the *Chronicle* identified them, included an attorney who represented her and a physician who treated her for an illness. In 1870 she brought a lawsuit against a merchant, somewhat resembling the one against Mackay, for $10,000 and costs. Six letters she received from a state senator were sold back to him for $1,000 so they would not be handed over to his wife (the original asking price had been $1,000 per letter). After one relationship with a married man soured, she went to the wife and offered herself as a witness to the husband's infidelity. She received a high-priced piano from a music dealer, jewels from a jeweler, and further tributes from other companions. It was reported that these were given "either voluntarily or otherwise."

Correlative information about Mrs. Smallman included the fact that the previous August she had been arrested for embezzling $2,000 from an Englishman in a mining-stock scam—although charges were dropped after half the money was returned. About the same time she avoided arrest after paying for a large quantity of perfume she had attempted to pilfer from a local drugstore.[31]

Owing to San Francisco's large number of men having reached sudden wealth, the "female adventurer" was said to have been a particularly prominent character in the city. The day after the Smallman litigation was announced, the *San Francisco Bulletin* printed an editorial estimating that "women in purple" ran blackmailing assessments in the city as high as half a million dollars a year. Despite the frequency of the felonious acts, the perpetrators almost always avoided legal action because of the notoriety that would result from prosecution.[32]

On the Comstock the *Territorial Enterprise* weighed in without mincing words. It attacked Mrs. Smallman, calling her shameless and notorious, and proposed that attorneys who represented such cases should be disbarred. The editorial began: "Another brazen strumpet, flaunting her depravity in the face of the public, has again instituted one of those blackmail proceedings for which San Francisco has become somewhat noted."[33]

A San Francisco detective related several schemes in which adventurers of the era conned unsuspecting victims. In one a man was lured to a woman's house when she asked to share his umbrella on a rainy day. Invited inside until the storm abated, the man was startled when the woman passionately threw her arms around him just before her husband rushed in to catch them. The man's watch and $150 secured a promise from the aggrieved husband to keep the incident quiet. A well-known lawyer was compromised after being left alone with the young daughter of a hostess on several visits. On the last visit the girl made advances toward the attorney, who could not resist. When "discovered" by the mother, he was only too glad to get off with paying $1,000. Women too could be vulnerable, as proved by a wealthy merchant's wife who was "decoyed into an assignation house, innocently, by a scoundrel who had professed great friendship." For four years she paid $25 a week to keep him from telling her husband.[34]

The direct cause of the filing of the Smallman suit was Mackay's refusal to meet with the Smallmans or answer a letter from the husband. The letter was dated November 29, 1878. Several days earlier Mackay had snubbed

Smallman and his wife when they came to meet him in Virginia City (the suggested meeting was close to, and may have been on, the Mackays' wedding anniversary—November 25). In the missive Smallman claimed that for some time he had received letters relating to a relationship between Amelia and Mackay. He at first attributed the missives to a jealous individual whose intention was to break up his happiness. Later, coming to believe the accusations true, he accused Mrs. Smallman. She had proposed the meeting so that Mackay would deny the relationship in front of Mr. Smallman. When Mackay refused to see the couple, Mr. Smallman claimed that Amelia had attempted suicide. Upon being saved by timely medical assistance, she confessed guilt in the illicit relationship.

The letter to Mackay continued that as an honorable man, the husband could not abandon his wife: she was now demented, without money, and in debt. Mr. Smallman wanted nothing for himself but believed a settlement for his wife was only just and would avert the indignity of a public case. Once her financial problems were resolved, he would quietly divorce her and end the dreadful affair. He concluded by saying that should he receive no answer within five days, he would be forced to begin action in the courts.

On December 10 the Smallmans' attorneys wrote saying that they were in the process of instituting legal proceedings and that if Mackay wished to settle the matter he needed to call at their offices immediately.[35] When the request was met with silence, the lawsuit was filed and the news hit the streets.

The case never reached the level of scandal it promised. As demonstrated by the newspapers' articles regarding the proliferation of adventuresses in the era, as well as the chronicling of Mrs. Smallman's notorious past, there was cause to suspect the charge fraudulent. The media had concluded that Mackay was a victim fighting back. The *Bulletin* seems to have been expressing public opinion when it concluded: "When a citizen fights one of these cases he renders the public a service." Under examination the accusation did indeed prove counterfeit. Amelia Smallman was found to be sane, and the case against Mackay was thrown out. The following spring the San Francisco grand jury brought thirteen indictments against the Smallmans for defrauding a carpenter out of $6,000 in a stock deal. The Mackay suit was used as evidence against them. Both husband and wife were convicted and sentenced to the California State Prison for four years.[36]

CHAPTER 10

To Try Fortune
No More

JOE STEWART was a friend of Mackay's from the Downieville days. He was cultured and distinguished in appearance, a man of impeccable manners. He was also one of the leading gamblers on the Comstock. In the late 1870s he needed money to keep his faro games going. On five occasions over a period of a few months, Mackay loaned him ten thousand dollars. Mackay never criticized or rebuked Stewart, but finally, while handing him a check, Mackay asked quietly, "Joe, w-what's your limit?" Stewart did not ask again.[1]

Mackay was erratic in his choice of friends: the gambler Stewart, the volatile Edward Stokes, and Fair. Old-timers on the Comstock wondered how the Mackay and Fair partnership lasted. Grant Smith commented: "The seemingly harmonious association of those two men of diametrically opposite characters, during the early years, was the subject of much comment among old-time Comstockers. The excuse made for Mackay was that foxy Jimmy did not display his true character until after he became rich."[2] The friendship failed with the extraction of the last great ore on the Comstock. There was no further need of a partnership, and so the friendship, strung together in the end by common goals, unraveled.

Despite his cynical reserve, Fair had natural leadership abilities and enjoyed not only the social status of a man of wealth, but also the stature associated with being a central figure in the Comstock community. He served on the Virginia City Relief Committee, and when Sam Curtis of the Justice Mine Company and the superintendent of the Alta mine were about to enter into all-out war, Fair led other superintendents in confronting the two and quelling the trouble.[3]

Fair had faith in his own judgment and knew Mackay was an exceptional miner, but he trusted few others. As years passed, Fair held partner Flood in disdain, because the San Franciscan could not possibly understand managing a mine while living "down below." Fair preferred superintending all the Firm's mines himself. When responsibility was delegated to others, even a superintendent such as Frank F. Osbiston who had years of experience managing mills and mines, Fair insisted on meeting once or twice daily to review decisions and issue directions. Fair toured every mine every day and got by on five hours of rest a night. The burden he undertook was immense, and he bragged about it, saying: "The receipts and expenditures of our Company total more than that of half the states in the Union."[4]

Fair paid a price for his constant attention to business. On June 1, 1878, the *Territorial Enterprise* reported that he was confined to his house by rheumatism: "His strong constitution is serving notices upon him that it cannot endure a steady physical strain upon it by day, and a steady mental strain day and night, any longer." During some illnesses in the past he had taken to his bed and remained there for long periods. His wife stated that sometimes, when fighting off a physical breakdown, he carried a quart bottle of brandy into their bedroom and "the bottle was empty the next morning." In the spring of 1878, whatever remedy he pursued did not affect a cure. In July, with Comstock ore running out and ill health continuing to plague him, Fair resigned as superintendent of the Firm's mines, retiring to San Francisco.[5]

Two months later he was back in Virginia City. At the north end of the Lode, beyond the Big Bonanza, the Ophir mine, and two others, was the

Sierra Nevada. Its president was Johnny Skae, who, along with Mackay and associates, was one of the directors of the water company. Skae had made a fortune while working as a telegraph operator, deciphering for his own use the ciphers sent between Flood and Mackay regarding the development of the Big Bonanza. By all accounts Skae had a genial and generous personality. On Sunday afternoon August 25, 1878, he hosted a party for 150 at a reservoir that he had stocked with thousands of fish. The guest list included every Comstocker of note. The entertainment included speeches and singing, drinking copiously, and fishing, which devolved after a time into attempts to grab the fish by hand. Alf Doten reported that it was as fun as any entertainment on the Pacific Coast. Skae had reason to celebrate: his Sierra Nevada mine had struck a pipe of ore, and he was following it downward. With production in the other mines faltering, increasingly desperate stock gamblers were speculating that the Sierra Nevada might hold the next great bonanza.[6]

Sierra Nevada stock had sold for $2.80 a share in June of 1878, but in early July, as the winze below the 2,000-foot level drove through low-grade ore, wishful investors began spending in anticipation of another boom. A short crosscut at 2,100 feet revealed a body of good ore, yielding one specimen that assayed at $900. The *San Francisco Bulletin* attempted to dampen the excitement, which was beginning to rival the intense winter of 1874–75. The paper reminded readers that the year the Con. Virginia had a sample of ore assay at $900, its average yield was $93, and that during the Ophir excitement, one sample that assayed at $27,000 merely created senseless excitement. But Dan DeQuille, who had been prophesying another strike in the north-end mines, praised developments. So did other mining reporters. A week after Skae's party, shares rose to $82. When bears drove the price back to $42, Skae ordered a crosscut at 2,150 feet, where prospects appeared particularly good. The *San Francisco Chronicle* reported that the cut exposed another first-class vein, with the ore in the face being particularly rich. As the winze reached 2,200 feet, a crosscut revealed 20 feet of ore, some of it very high grade. At that point Skae ordered the work stopped. The vein seemed to be leading into the adjoining Union mine, and shares of Union stock rose dramatically. In sympathy all Comstock shares increased.[7]

The market was unstable throughout September, but prices generally rose. Sierra Nevada stock fluctuated from $220 to $161 to $210. The *Gold Hill News* consistently reported that the ore on the bottom of the Sierra Nevada was richer than ever. Flood and Fair were among the buyers of Sierra Nevada and Union mine shares. On September 26 Fair arrived on the Comstock to get a firsthand view. The price of the Sierra Nevada stock was approaching $280, and Union was selling at $182. It was the latter, toward which the vein dipped, that most interested Fair. The president of the Union was Robert Sherwood, a friend of Flood's. Sherwood had sold Flood 1,000 shares of Con. Virginia stock for $100,000 shortly before the price doubled in 1874. Flood later remarked to Sherwood: "We built [the Nevada Block] from the profits on that 1,000 shares."[8]

Fair sought out Sherwood, and they went into the Union near the Sierra Nevada boundary line. Fair made a thorough inspection and immediately offered Sherwood $200 a share for his 5,000 shares. Sherwood balked. "The offer is not a continuing one," said Fair. "That's a million dollars and I will give you a check for the money right away if you want to sell." Sherwood took the offer. Fair, returning to San Francisco, stopped by broker George Mayre's office and told of the purchase, perhaps hoping to bull the market by letting out word. Mayre only commented that he could have gotten the stock cheaper.[9]

Between the shares they already owned and Flood's purchase, the Firm now owned the majority interest in the Union mine. Over the next month the market fell. Mackay came from Paris directly to Virginia City to examine the new strike, arriving on October 18. He saw Sierra Nevada's low-grade ore between the 2,000-foot and 2,200-foot levels and three short crosscuts, each with pockets of rich ore but the longest being only 20 feet. There was no proof of extensive bodies of ore, as had comprised the big bonanza in the Con. Virginia. The Union mine revealed even less. Mackay used their code to telegraph Flood, "Fair is crazy." Sherwood afterward told Flood: "I built the Union Block [in San Francisco] with the profits on that 5,000 shares."[10]

The steadily declining prices emphasized the foolhardy nature of Fair's purchase. The day after Mackay inspected the mines and while he was still in Virginia City, Fair left San Francisco for the East.

On October 22 the leading stock, Sierra Nevada, was selling for $165. During the next month, the market rallied. When it opened on Monday, November 18, Sierra Nevada was back up to $200 per share.

On the Comstock, on the evening of November 18, there was to be a serenade to favorite son Johnny Skae. The largest band ever assembled in

Nevada, seventy-five musicians, had been engaged. But the event never took place. For some time a group of owners had realized that the Sierra Nevada and the Union were not developing as anticipated. They had been secretly selling large blocks of stock. On the weekend of November 16 and 17, a rumor circulated that the Firm's expert, Frank Osbiston, had been denied access to inspect the Sierra Nevada. Insiders suddenly lost confidence. The morning of the 18th, brokers panicked, and Sierra Nevada broke to below $100. Never before had a stock leader lost half its worth in one day. All other stocks were carried with it. With the crash, stockholders in Virginia City created a riotous scene. Skae had made a substantial amount of money selling stocks. Some believed him to be involved in the market break. Others disagreed, pointing to his refusal to allow outsiders to view the mine as an attempt to prevent it. Fearing violence against Skae, special deputies were quickly sworn in. It proved unnecessary, because of the ruined investors' inability to agree as to Skae's involvement. A deputation from the Miners' Union examined the mine that afternoon and reported good prospects, but the bubble had burst. The next day shares of Sierra Nevada stock fell to $65.[11]

Two days later Alf Doten commented in his diary: "Stox busted wider today than ever before." The *Gold Hill News* reported a "deep-seated agony" and "pale, woe-stricken faces" as crowds gathered around bulletin boards that recorded the destruction. "[They] watched the hard earnings of years of savings and toil go like the drifting sands beneath their feet. . . . Tongue cannot describe the ruin." The devastation continued day after day. Although six months later, in June 1879, the Union mine rallied to $99 a share, it was another illusory flash, soon falling to $26½.[12]

The ore in the most promising of Skae's Sierra Nevada crosscuts did not continue far. It was famously dubbed the "coyote hole." Some said Johnny Skae was the only one to come out on top; others reported that he went down with the crash. As regarded Skae's fortunes, it mattered little which account was correct. When in the money he lived recklessly, reportedly losing $60,000 in one game of poker at the Palace Hotel. He came to a pitiable end. On August 25, 1881, the *New York Times* reported: "Many old Californians shook their heads thoughtfully yesterday morning over their coffee as they read in the records of the city prison the condensed testimony of 'Johnny' Skae's extreme poverty." He died four years later, penniless to the end.[13]

Grant Smith remarked that the Sierra Nevada stock deal broke the hearts of the people of the Comstock. They felt they had been duped, and it seemed their last chance—the crash left them hopeless. In San Francisco the market never recovered. The Firm, left to make what it could from the small bodies of ore in those mines, milled 83,836 tons of ore from the Union, Sierra Nevada, and Ophir mines, about 10 percent of the amount they had milled from the Con. Virginia. Although $5 million was taken from two adjoining ore bodies between 2,200 feet and 2,500 feet, expenses left nothing to be paid in dividends.[14]

As early as February 1877, with only the bonanza mines working in good ore, unemployment became the bane of the territory. The *Virginia Evening Chronicle* had commented at the time: "The hard times are beginning to tell on everybody, high and low." By 1879, with even the bonanza mines in decline, residents in droves abandoned the town. Perhaps because little good ore was being worked that summer, mine managers began a policy of not admitting the public underground.[15]

On the evening of August 4, Mackay and Belcher mine superintendent W. H. Smith were sitting in a livery stable when Miles Finlen, a coal dealer and notorious professional fighter, approached. As Smith talked to someone else, Finlen—who later killed a man in a fight—demanded that he be allowed into the Sierra Nevada mine. Mackay answered that no one was allowed in, saying that those permitted to enter had abused him. Finlen objected and moved toward Mackay. A bystander started to step in, but Mackay said: "Let that damned coward come on—I'll take care of him." One of the two, no one reported which, threw a punch, and they grappled on the dirt floor. Separated, Finlen went away vowing vengeance. Later Finlen, saying Mackay had called him names, went about the streets with two revolvers, acting wild. Mackay attended Piper's Opera House with friends. The two did not meet that night, and mutual friends kept watch thereafter, so the battle was never rejoined.[16]

In November 1880, under the recently revised California Constitution, which sought to alleviate the excessive tax burden of farmers and increase the responsibility of bank and corporation directors, Mackay was assessed for $36,000,000 in personal property and $250,000 in cash. His stock holdings included extensive shares in the Nevada Bank of San Francisco, the Pacific Mill and Mining Company, the Pacific Wood, Lumber and Flume Company, San Francisco Gaslight, Virginia and Gold Hill Water Company, a chemical works, two dynamite companies, and the Ophir mine. Although the *Territorial Enterprise* termed the Mackay assessment and others strange, the appraisal showed that he was one of the richest men in the world.[17]

That fall Fair returned to the Comstock from a seven-month voyage around the world. A delegation greeted him in royal fashion. The welcome included shrieking mine steam whistles, bunting and flags decorating many of the buildings in town, and a serenade at his house. While gone, he had agreed to run as a Democrat against the unpopular incumbent, Sharon, for the seat in the U.S. Senate. Fair's run culminated in his election on January 12, 1881. He owed the victory to the fact that his operatives carried sacks of cash around openly buying votes in the legislature, which elected the U.S. senators. Sharon associate Henry M. Yerington said: "Fair spent about $40,000 in Washoe and Eureka counties." The *Gold Hill News,* supporting Sharon, called it a miserable defeat. "The trouble here in Nevada is very well understood and appreciated," said editor Doten. "Colonel Jim Fair and his friends were too well supplied with the munitions of war. He literally sacked the state."[18]

Mackay was in Virginia City during parts of the campaign. When he arrived on November 12, 1880, the *Territorial Enterprise* announced: "John Mackay returned from San Francisco yesterday morning and at once dropped into his working harness." A short time later he was involved in an incident in support of Fair. Sharon had recently hired a brilliant newsman, Fred Hart, as editor of the *Enterprise.* A Hart editorial entitled "Slippery Jim," induced as Doten recalled by "too much brewery inspiration," spent twenty-one paragraphs ridiculing Fair. When Mackay saw the item, he marched to the paper's offices in something approaching an apoplectic state. "I own one-half of this cussed paper myself and won't have my partner Fair abused and belied in it by anybody. I've a d—— good mind to take a sledge and smash h—l out of the bloody press." Hart was said to have survived by judiciously staying out of Mackay's way.

Doten reported that within a few weeks Hart again fell under the influence of the east wind from the brewery. This time the editor took up the cause of the Justice Mine Company, which he contended had been robbed by those associated with the Alta mine. Whether this was a continuation of the feud between the adjoining mines of two years earlier is unclear. But the accusation impugned Mackay and a group of his associates who now controlled the Alta. Doten says Mackay and friends swarmed the *Enterprise* office like hornets. Getting wind of their intent, Hart had already made "a lively break" down the road and over Geiger Grade—continuing away from the Comstock until obtaining employment in San Francisco.[19]

The incidents were again brought to public attention eighteen months later, when the widely admired Father Manogue was promoted to bishop. The *Enterprise* commented offhandedly that the prelate's influence had helped him acquire the position. When the new bishop took exception, saying he had earned his promotion, he spoke of Mackay's threatened use of a sledgehammer against the press. At that the *Enterprise* took exception, saying Bishop Manogue's comments reflected on management even though Hart had long since deserted his editor's post. The paper explained that when Fair had been lampooned in a light manner by Hart, "Mr. Mackay thereupon waited upon the latter and told him that Mr. Fair was his partner, his associate and friend; and, while it was the editor's duty to oppose his candidacy, he objected most strenuously to any personal villification [*sic*] of the candidate. . . . There was no violence, no sledge, nor any coercion whatever." It concluded that Mr. Mackay had been "grossly misrepresented by the street fiction."[20] But with the street version publicized by the territory's leading cleric a year and a half after its purported occurrence and by leading Comstock journalist Doten years afterward, it became part of the Mackay legend.

On December 1, 1880, Mackay arrived in New York with Rollin Daggett. The newsman was now representing Nevada in the House of Representatives. When Mackay reached his hotel wearing a rough tweed suit and a dusty slouch hat, it caused amused comment among the stylish guests who were present. He adjourned to his rooms, emerging a time later in quite different attire for a dining engagement with ex-president Grant. A reporter for the *New York Graphic* was frustrated in attempting to interview him, commenting that Mackay was not very talkative. When pressed about the condition of the mines, Mackay spoke softly, slipping at times into a slight Irish accent: "I am hopeful, and that is all I care to say. . . . Were I to talk much I would be misconstrued and blamed, as I have been so often before. Time will tell all."

Daggett told the reporter that, having talked with Mackay, he knew the bonanza king had unlimited confidence in the mines—both reworking low-grade ore and deep mining. In speaking of Mackay, Daggett said he was one of the most unselfish men he had ever met, having given away twice that of his partners, who were themselves exceedingly generous. Mackay, he concluded, mixed with all classes and "would any day rather take an old miner by the hand than a millionaire."[21]

Mackay traveled from New York to Paris, and just before Christmas 1880 the Count and Countess Telfener hosted John and Louise Mackay at a dinner in John's honor. The Telfeners' son and Louise's younger sister, Ada, had been married in March 1879—a wedding John had not been able to attend. On that occasion Ada had written John, "I can only say one thing that will give you an idea of his character, he is a second John Mackay." At the Telfener party, singers from the opera entertained. The elegant appointments, including a fountain featuring goldfish and pearly tropical shells as the table centerpiece, created an atmosphere said to resemble something from the pages of the Arabian Nights. The *Territorial Enterprise* added its comment to a lengthy description of the event: "No doubt Mr. Mackay would all the time have been better satisfied to have been on the 2600 level of the Union Consolidated."[22] Three weeks later a sign of the times was on display on the Comstock: the main building of the Con. Virginia shaft was dismantled and rebuilt over that of the Hale & Norcross.[23]

At the end of February 1881, Louise hosted a ball that was the most talked about event of the season. Queen Isabella of Spain was understood to have been among the four hundred guests, although at least one newspaper account contradicted the report, saying there had been a rupture between the queen and the hostess and they had not visited together for some time. John presented Louise a chain of diamonds that evening that she wore coiled from the curls on the crown of her head, circling her neck several times and falling in wide loops to her waist. Singers from the opera and actors from the Comédie Française entertained in the ballroom, while a temporary roof was erected over one of the gardens to provide a second ballroom for dancing. Eight gilded columns supported the roof, eight immense mirrors with garnet velvet hangings alternated with tapestries on the walls, and a dozen great chandeliers illumed the room. Dinner was served at midnight, followed by suppers at 2 AM and again at 4. John Mackay was not in attendance for the suppers, having retired earlier. Newspaper accounts agreed that it was a magnificent entertainment and that it would be difficult for any host or hostess to equal it. Louise was on her way to becoming the most talked about social personality ever to come from America's West Coast.[24]

The next great social event, in early May 1881, brought together guests from every European capital. It was a reception and ball in the Mackays' honor held at the estate of the host, James Gordon Bennett Jr., the son of Mackay's boyhood hero. Bennett had taken over the operation of the *New York Herald* newspaper after his father's death in 1872. For the ball private express trains were hired to transport the guests from various cities. The morning after the ball Mackay and Bennett walked in the estate garden, immersed in a discussion that would lead to Mackay's next great venture.[25]

The Mackays were scheduled to travel to London to stay at the Grand Hotel Charing Cross in May, but the assassination of Alexander II in Russia on March 15 caused President Chester Arthur to ask Mackay to serve as a special ambassador and attend the coronation of Alexander III.[26] Although rooms in Moscow were impossible to secure, the Mackays occupied the first floor of the Hotel Dessaux. Soon after their arrival they gave a dinner for the members of the U.S. Special Mission. On May 15 they were presented at the Russian court. The empress was especially gracious, chatting pleasantly about California. The Cathedral of the Assumption in Moscow was as grandiose as anything in Rome, and the reception and balls were splendid, but Ellin Berlin said Louise Mackay came away from it disturbed. The long journey seemed across centuries as well as miles. Louise was uneasy about the emperor and his guests, who believed they were a ruling class with total, unlimited power. Alexander III, who soon represented himself as chosen to defend autocratic authority, became known for his persecution of racial and religious minorities. Although continuously guarded by police agents, he would die at the age of fifty, worn down by fears that revolutionaries would assassinate him as they had his father.[27]

Returning to Europe, Mackay spent several days in Rome and ended his stay in England. As was to become his habit, some of his tour was used to buy art. A year later, in 1882, Mackay commissioned one of the most famous modern artists, E. Meissonier, to paint a portrait of Louise. The result was unsatisfactory to the sitter—the painting made her look too old, and the hands were taken from someone else. At first she refused to accept the painting, recommending that Mackay not pay. French papers criticized her for the quarrel, and in the end Mackay paid the $25,000 fee.[28]

Exaggerated reports of his wealth, in some newspapers comparing it to the stories of "Ali Baba and the Forty Thieves" or "Aladdin's Lamp," brought letters imploring financial assistance from around the world. They were addressed curiously: "Mac-kay, esq., Bancker, Nevada or New York"; "Mr. J. W. mackay, Esquire, possessor of the silver mines of Great Bonanza, Nevada, California, America"; or simply, "Mr. Mackay, North America."[29]

Mackay returned to Virginia City at the end of June 1881, but he stayed only a short time. The riches in Nevada's mines continued to dwindle, requiring less and less of his time. In San Francisco bickering was rumored to have begun between Nevada Bank president Louis McLane and Flood and Fair. In October 1881 McLane retired, and Flood assumed the position of bank president.[30]

In 1877 the Comstock district had produced $37,062,252 in ore, down $1 million from the all-time high the previous year. In 1878 production fell to $20,436,685, and in 1879 to under $7,500,000. Deep mining, spearheaded by Mackay, began in earnest in 1878 but yielded little. The California mine paid its last dividend in 1879, the Con. Virginia its last in 1880. They were the last two mines paying dividends. The total production of the entire Comstock Lode in 1881 would barely total $1,000,000. In the papers the list of unpaid assessments grew longer. In 1881 Eliot Lord concluded: "The time may come when even such daring speculators as Mackey [sic] and Fair must count the cost of their ventures carefully and decide to try fortune no more."[31]

Until 1883 Mackay and Flood, but not Fair, continued to invest in deep-mine exploration. In 1876 they acquired control of the Yellow Jacket and sank a new shaft to 3,060 feet, an investment of $2,000,000 that yielded no ore. In October 1880 Mackay expressed a certainty that beneath the richest bonanzas more veins would be discovered, and he invested in that belief. He undoubtedly was influenced by Ferdinand Bacon Richthofen's classic work of 1865, which concluded that although the richness of ore bodies would not equal those of the surface bonanzas, the ore-bearing character of the lode and the continuity of the vein "must be assumed as fact," and the 1882 report of George F. Becker of the U.S. Geological Survey, who wrote that with the mines now six times as deep as on the date Richthofen examined them, "his opinions and predictions have been for the most part verified in the most remarkable manner." Mackay asserted: "Although we may not get these bonanzas in the next five years, still they will be found." The North End mines: the Ophir, Mexican, Union, and Sierra Nevada, were operated as if they were one mine—without result, even though the Ophir-Mexican shaft reached 3,360 feet. Shafts were sunk beneath the Con. Virginia and California veins, likewise losing the millions of dollars invested.[32]

When J. P. Jones showed that mining low-grade ore from old stopes could be profitable, Mackay enlisted Jones and Flood's son, James L. Flood, in a partnership that reworked the old mines between the years 1884 and 1895. Two skillful superintendents, W. H. Patton and D. B. Lyman, led the recovery efforts, and over that eleven-year period the mines yielded coin value of over $16,000,000. A friend of Mackay's said that Mackay guessed as much money had been expended in the hopes of finding new ore as had been taken out. In the end Mackay was forced to change his opinion regarding bonanzas at depth. The friend reported hearing Mackay say that theirs was the last.[33]

Most miners left the Comstock as the rich ore died out in 1879 and 1880. In the next ten years fully one-fourth of Nevada's citizens moved out of the state. The area's malaise was typified by an encounter between a young Grant Smith and Mackay in 1887. Smith, desperate for work and having applied to the mine superintendents without result, intruded on Mackay at his room in the International Hotel. Mackay had been warring over new enterprises and was in the process of saving the Nevada Bank of San Francisco, teetering on the brink of collapse. Smith reported that Mackay looked old and worn. He was shaving, in his undershirt, his chest covered with white hair. Smith asked for a job, saying he needed it badly. Mackay replied, " 'I have nothing for you,' and the door closed."[34]

In October 1895 Mackay visited Virginia City for the last time. Those still speculating on the mines regarded his visit as a positive sign. He spent a day with Superintendent Lyman examining the lower levels of the Con. Virginia's south end. But he saw nothing encouraging. When he left, Mackay—who, with Jones, had been the only man of wealth attempting to keep the industry alive—abandoned the Comstock for good.[35]

CHAPTER 11

A Changing Cast

JAMES FAIR'S career in the U.S. Senate was more distraction than vocation. He found the role dull. He enjoyed being addressed as senator and pontificating on matters of the day, but he paid little attention to legislative matters. It was said that he spent much of his time in Washington in his office visiting with other malingering lawmakers or entertaining guests from the West, a bottle of brandy close at hand. He gave one speech on the Senate floor in his six-year stint: having promised voters he would address the issue, he spoke in support of the Chinese Exclusion Act of 1881.[1]

With politics not as he had imagined, Fair spent much of his time in San Francisco rather than the capital city. The mines were depressed, and he thought investing in deep mining a waste of money, so he withdrew from his partnership with Mackay and Flood. He had bought a large share of a Bay Area railroad system, which he sold for $500,000 in cash and $5,500,000 in bonds, with interest guaranteed by the Southern Pacific Railroad Company.

He was the only one of the investors who made money on the railroad, but a partner in the venture commented: "I don't think [Fair] took much interest in the narrow-gauge road further than to assure himself that his money was being well invested. . . . He didn't know much about railroad matters and cared to know less. Fair was a miner." The career to which he was best suited was no longer available, so he consumed his days investing in real estate. Evenings, in the finely appointed bar of the Lick House, he regaled listeners with tales of the Comstock. Withal, he ate ravenously, drank, and chased women.[2]

Between 1878 and 1888 the number of divorces in America swelled from sixteen thousand to twenty-eight thousand. In twice as many cases, the woman was the complainant. Owing to reform movements, social interests outside the home, and increased employment, women were no longer suffering ignominies or ill treatment in silence. In 1883 Theresa Fair accused Senator Fair of infidelity. Once accused, Fair feigned suspicion of her. He became morose. The *San Francisco Examiner* later reported that he was "even murderously inclined," and "[he] pretended to believe things were absolutely impossible. Twice he attacked her in the presence of her children." Mrs. Fair was a devout Catholic, and Bishop Manogue was called upon to try to prevent a separation. But Theresa feared for her safety and dared not stay in the same house with Fair.[3]

In the May 8, 1883, *New York Times,* the headline "Senator Fair Sued For Divorce" was page-one news. Mrs. Fair was charging him with adultery. Two specific San Francisco women were named as having aided his faithlessness, their addresses given. The first incident had occurred in November 1881, the second on April 27, 1883. Fair chose not to contest the charges. He portrayed himself as protecting his wife while implying there was more to the story than his moral principles would permit him to divulge. The *Times* stated: "[Senator Fair] regretted bitterly and keenly the notoriety given a matter so exclusively concerning himself and wife. He said: 'I am the man, and am willing to bear all the odium which the public, in its ignorance of the facts, may choose to cast upon me, but my regret is for my wife, whose name should not be associated with and incorporated in dispatches transmitted all over the country.'"[4]

During the trial, one of the women named in the complaint testified that Fair "participated in carnal acts as alleged." The other gave a deposition that also supported the plaintiff. Theresa Fair, renowned for her charities, was treated in courteous fashion by the papers. Being both unpopular and guilty, James Fair was not. Two years afterward the *St. Louis Globe-Democrat* commented: "Public opinion was against him at the time of the divorce suit of his wife, and nothing has since changed it."[5]

Mackay, a friend to Theresa Fair since she had helped bring Louise and him together, stood by her during the ordeal. This caused a final breach between Mackay and Fair. Mackay advised Theresa, and she hired Judge R. S. Resnick, from Virginia City, as her attorney. The result was that she was granted an uncommonly generous award for the era: one-third of her husband's fortune, the newly acquired home on Pine Street in San Francisco, and custody of their two girls. Fair was given custody of their two sons, and his contention that he needed two-thirds of the fortune to take care of the boys was upheld.[6]

Because of Resnick's involvement, Fair announced he was through with the Comstock. He became bitter toward Mackay and Flood, who had not supported him either. After one of his interviews with Fair, George Howard Morrison, a biographer for historian H. H. Bancroft, said: "[Fair] don't feel kindly toward his associates. . . . When he gets right down to talking with me confidentially he besmirched every one of them." Mackay was the leading target. Morrison illustrated how Fair disparaged Mackay in offhanded comments. Fair told him: "I don't think Mackay ever had charge of a mine—he is not a man for that kind of business. I would not care to say anything about that though." In public Fair's enmity often surfaced in quips. When Mackay gave an elaborate dinner for the actor John McCullough, Fair commented to the papers: "John Mackay's a great admirer of geniuses. I wonder how much they borrowed from him."

Flood backed Mackay in any dispute with Fair, and Fair—already disdainful of his banking partner's mining aptitude—directed scorn toward him as well. When Flood began building an ornate home among wealthy former southerners in Menlo Park, Fair made certain Flood's chivalric neighbors were aware of his background. He remarked to the press: "Flood should be popular at Menlo. There's not a bartender on the Coast who can make a better julep than Jim."[7]

For over a year Fair's dealings in Washington, D.C., and in San Francisco real estate, along with Mackay's new business interests in New York, kept the men apart. Fair had withdrawn from the Firm's mining interests, so their only partnership was in the Nevada Bank, which since its inception had been one of the largest financial institutions in the West. It was well regarded across the country, as evinced by the *St Louis Globe-Democrat*'s describing it as "a little mint in revenue." When O'Brien died in 1878, he left his bank shares to his two sisters. Women were not eligible to serve on a board of directors in the era, so they sold the shares to Flood, Mackay, and Fair. When Louis McLane retired in 1881, he sold his shares to the trio as well. Experienced vice president George L. Brander was sold a nominal amount of shares, as was Flood's son J. L. Flood, so they might serve on the five-person board required by the bank's bylaws.[8]

In the summer of 1884 the differences between Mackay and Flood on one side and Fair on the other became pronounced. Mackay and Flood agreed that they had to sever the relationship. They told Fair that if a settlement was not reached, they were prepared to dissolve the corporation. They offered to buy his shares or have him buy theirs. Fair said he would take over full ownership, but when the papers were drawn he backed out.

Mrs. Mackay had moved to London, and Mackay had business on both sides of the Atlantic, including his stepdaughter Eva's wedding in February 1885 in Paris. In November he left for New York, preparing to sail for Liverpool. Flood and Fair met on November 19, and Fair agreed to sell his shares for two million dollars if Flood did the same. If Mackay agreed, he would become the full owner. The proposal was a ploy by Fair, who believed Mackay's heavy investments and business reversals did not leave him with enough cash to afford the purchase. Telegraphed the proposal, however, Mackay immediately accepted. On November 22, 1884, Fair conveyed his shares of stock to Flood, who—as Mackay's agent—was acting for him. In the agreement Flood was to continue his roles as bank president and a director for one year.[9]

Eva was wed to Italian Don Ferdinand Colonna, the Prince of Galatro, in a Pontifical High Mass performed by the Papal Nuncio in Paris. The papers touted the reception, ranking it with the most magnificent festivals in all of France. Mackay had apparently become more comfortable at Louise's fetes. England's *Pall Mall Gazette* commented, "The Big Bonanza was perfectly at home among the princes representing the oldest houses in Italy, and anyone who knows how to read a face would say that he was a man to whom they should all take off their hats."[10]

In May 1885, after Mackay returned to America, word surfaced that Flood was buying back his interest in the bank. An article in the *St. Louis Globe-*

Democrat that month identified Mackay as popular with a public confident in his integrity; the same article labeled Fair as "the sharpest and most unscrupulous of all the bonanza millionaires." Days earlier Fair had been interviewed in the *San Francisco Chronicle*. He said that when Flood approached him about selling bank shares, Flood's health had been bad. He wanted out of the bank so he would not "die in the harness." Fair listed the reasons he did not accept the bank's sole ownership: he was not satisfied with the way the bank had been managed; he did not want a sedentary occupation; and he did not want to die in the harness himself. When asked if rumors might be true that Flood was rejoining his partner Mackay in the enterprise, Fair said it would not surprise him. Did Flood's reinvestment mean Fair had simply been frozen out of the association, as was now common speculation? He would not comment other than to say that his resignation had been perfectly voluntary and well considered. Asked if it were a Machiavellian transaction, quick witted as always, Fair replied that it looked "like a Floodavillian deal."[11] At the annual meeting of bank stockholders in October 1885, Flood was reelected to the board, net earnings of $504,000 were announced, and a dividend of 10 percent was declared.[12]

The reason Fair did not believe Mackay would be able to buy the others' interest in the bank the year before was that Mackay had begun a momentous new enterprise. In August 1883 he had decided to challenge the communication monopoly of the legendary financier Jay Gould. Mackay and his partner, James Gordon Bennett Jr., the owner of the *New York Herald*, had declared war on Gould's trans-Atlantic cable monopoly. Because it was on a world stage, involving international communication and three of the era's biggest names, this financial war was to be more spectacular than Mackay's Comstock battles.

Mackay had not been ready to retire when the mines fell on hard times. He seemed able to abide only short periods of Louise's life of leisure in stately society. In the 1880s, when state legislatures named U.S. senators, he had every opportunity to become involved in political affairs. In 1883 he told a reporter: "I am not fool enough to go into politics." When told the following year that his name had been continually mentioned as a future senator, he replied: "I know it has, but nothing could be more absurd or further from the truth. I have no taste for political life, and nothing could induce me to enter it." In 1885 it was proposed that he be nominated in Nevada under circumstances that virtually guaranteed his election, but he turned it down. Still, he needed challenges and competition. In speaking of playing poker in the days when he had little money, he commented: "If I won I liked it, and if I lost I felt it; but now I don't care whether I win or lose. I have lost the taste for it entirely, and, candidly, I miss it." Financial warfare became his poker game.[13]

In the middle of the nineteenth century, technological innovation accelerated, notably in transportation and communication. Revolutionary rail and telegraph systems expanded capabilities in distribution and oversight. Coordination in the flow of goods, along with increasing consumer demand, led to an economic, social, and political transformation in America. Visionary capitalists built empires, changing the agrarian society of small businesses to an urbanized, industrialized nation. While the new corporations provided the benefits of efficient use of resources and lower prices on goods, they also caused alarm. Financial power was becoming concentrated, and long-established firms were overrun or usurped. Perceptions and values shifted, and belief in competition, business ownership, producing, and saving were challenged. The nation had begun its transformation from a land of individual producers to one of consumers.[14]

As an immigrant of nearly unrivaled success, Mackay felt an intense gratitude to the United States. He told essayist John Russell Young that he believed his wealth was merely a trust—that he owed his country its use in some special service. Mackay deliberated building a merchant fleet of American steamships that would compete with the great European fleets. He felt it would be part of the return he thought he owed. But before embarking on the new enterprise, while in France with Louise, Mackay met Bennett and rethought his second career. The line of enterprise he chose was influenced by the belief that power in the hands of the very few was threatening.[15]

When James Gordon Bennett Sr. died in 1872, his son, James Gordon Bennett Jr., had been running the *New York Herald* for five years. Still, it was widely assumed the paper of the masses would fail upon the father's death. Bennett Sr. had built it into a remarkable success. The son had won a reputation in the field of sports but was not thought to have much future in the news business. Contrary to public estimation, his efforts not only sustained the paper's place in American journalism by expanding European offices and printing European editions, it broadened its scope, giving it international import. Bennett Jr. had a grasp of world affairs and a picturesque way to present them. He was enterprising and open minded, courteous, frank and implacable.[16] He also was a fractious eccentric.

James Gordon Bennett Jr.'s school years had been spent in Europe. His father's frequent newspaper attacks against the powerful had earned him many enemies. His mother, not wishing the children to suffer, took them to live in France. Bennett Jr. acquired French but grew into a strange, ill-mannered young man who never acquired his father's skill with a pen. Although his father was hated by most of New York's social set, upon his return the son—the third wealthiest of the city's elite after young Vanderbilt and Astor—was accepted.

Bennett Jr. was renowned for his achievements in yachting. He had served in the navy in the Civil War, was the youngest commodore ever in the New York Yacht Club, and won the first transoceanic boat race. Unfortunately, he was also infamous for drunken escapades. There were stories of a wild nighttime ride, stark naked atop the coach; walking between tables of a restaurant pulling tablecloths, silver, and china to the floor; and, when a fire broke out at Delmonico's, exiting the restaurant in his evening wear and assuming direction of the firemen until they turned the hose on him. Finally, his career as an ex-patriot was secured on New Year's Day 1877. After visits to several homes, where punch bowls were kept filled, Bennett arrived at his fiancée's father's house, where, as his biographer Don C. Seitz writes, he "became guilty of conduct unbecoming a gentleman—or any one else." He urinated into the drawing-room fireplace.[17]

Two days later the woman's brother beat Bennett with a cowhide, after which Bennett challenged the brother to a duel (the challenge was carried by Charles Longfellow, son of the poet). Each man announced himself satisfied after firing and missing, and Bennett left the scandal and disparagement of New York for France.[18]

Several years earlier Bennett had told Henry M. Stanley that Bennett's father had made the *Herald* great and he intended to make it greater, no matter the cost. With that he had sent Stanley on his journey to find Livingstone in what was then called the "Dark Continent." After his exile Bennett funded a tragic expedition to the North Pole. The yacht *Jeanette*, named for Bennett's sister, was sunk by ice, resulting in the death of the commander and most of the company. Bennett later paid Dr. Frederick Cook twenty-five thousand dollars for his never widely believed story of finding the North Pole. The newsman's support of these and other explorations resulted in the naming of Bennett Mountain and Bennett River in Africa, Bennett Island in the Arctic, and Bennett Lake in Alaska.[19]

When Mackay and Bennett met in France, the *New York Herald* had been, for a number of years, as powerful and profitable as any paper in the world. But newspapers like the *Herald*, transmitting news from Europe, were laboring under exorbitant rates charged by the cable companies. The most conspicuous example was the *New York Tribune*, spending five thousand dollars to transmit a single dispatch on the Franco-Prussian War in 1870.

In 1858, when the first successful Atlantic cable was laid, financed by the British, people thought it a miracle that Europe and America were united by telegraph. Communication that had taken weeks now arrived in minutes. News of the cable was a sensation even on America's West Coast. In San Francisco a holiday was declared—although it was a month after the fact, since news of it had to be brought from the East by Panama steamer. When long-distance submarine telegraphy became commercially viable with the completion of two cables in 1866, rates were prohibitive: $100 for a ten-word message. Rates, later cut to $46.80 for ten words, were accepted with the rationale that trans-Atlantic communication was a luxury and those who used it could afford it. The rates fell when a competing line, laid by the French company the Société du Cable Trans-Atlantique Français, was completed in 1869.[20]

By the early 1880s rates were 75¢ a word, still excessive, especially for those who wished to send frequent or lengthy messages. There were eight cables between the two continents, but they had formed a pool that regulated prices. Bennett managed his paper from Paris and used more Atlantic dispatches than anyone else. He appealed to Jay Gould, who had been granted exclusive rights to control the cables, asking a discount for the *Herald* news agency. Gould, who owned or had confederates working in several other newspapers and controlled Western Union Telegraph and the Associated Press, granting them special rates, flatly refused.[21]

Mackay harbored his own antipathy toward the cable companies, because of the considerable sums he spent sending messages one way or the other across the ocean. Although his support of labor and the downtrodden reflected a progressive bent, Mackay was not anti–big business or even antimonopoly. He had run the most profitable mining operation in the world and understood owners' utilizing economic muscle to protect and expand their enterprises. The public, beginning to protest the growing number of business combinations and certain bankers and corporation managers who wished to maintain rate structures, were looking for federal intervention.[22] Mackay had seen enough of political machinations during the mining-tax brouhaha in Nevada to be suspicious of state controls. In this instance Gould's monopolistic behavior was a public detriment that affected him, and it moved him to act.

In February 1881 the *New York Times* editorialized about Gould and the First Amendment: "To the control of the telegraph lines Mr. Gould is now said to be determined to add control of the Associated Press, and it is popularly believed that by the purchase of one more newspaper he will achieve this end. He will thus, to a large extent, control what has been the free press of America." Giant enterprises were creating change in American life as local, family-run businesses were replaced. Here was a case of big business's seeming to threaten democracy itself. Mackay decided upon the service he would to do for his country: he would attempt to break the Atlantic cable cartel.[23] Mackay's ally, Bennett, saw himself as a reformer. His father had battled the telegraph monopoly in the 1860s. This crusade against Gould and the cable industry would be the son's extension of that fight.

One of the issues in the current situation was that the trans-Atlantic cables were built and run using large amounts of foreign capital. In times of emergency or war the British and French might dominate use of the cables, over which they held financial control. Mackay planned to enlist other large users to invest, and they would lay an all-American cable. Ominously, no one could be convinced to join in the battle against Gould.[24]

It is of some use when writing biography to judge an individual's moral standards by those of his or her era, while evaluating the era's standards against modern concepts. Jay Gould's financial career involved much the same behavior as that of other Gilded Age capitalists, but unique qualities of vision, daring, and ruthlessness enabled him to become the most powerful of all Wall Street speculators. Gould's actions included insider trading, bilking investors, buying government officials, bribing judges, and using the courts not only to attack, but also to obfuscate and delay. He was notorious to the public, while the era's sober businessmen viewed him as an architect of disorder. He was called among innumerable variations: "this ghoul in human form," "the Little Wizard," and "The Skunk of Wall Street." Competitor Daniel Drew famously said: "His touch is death."[25]

Much of the Gould legend developed because of his astute business sense. As corporations grew in size and diversity, managers became increasingly vital to operations and separated from owners. Gould was quick to grasp the new relationship between management and ownership. Although closely scrutinizing his operations' chiefs and bombarding them with directives and suggestions, he also gave them authority and encouraged them to use their initiative to run his businesses efficiently. Unless competitors challenged the operations, he generally let the career managers make the decisions. Those below the level of company superintendent were treated quite differently. They were depersonalized, regarded as mere pieces of the organization. Because his expanding enterprises required great amounts of capital and because extra costs, not price, had become the primary basis for competition, Gould continually fought to reduce operating expenses, including workers' salaries.[26]

In 1874 Gould controlled seven coal mines in Wyoming. When hard times and low wages drove miners to attempt to unionize and strike, he replaced them with Chinese workers—thereby reducing wages from $52 a month per man to $32.50. Reducing the labor expenditure cut the cost of production 25¢ per ton. At the time he wanted to cut it another 25¢, but he deferred to his western adjunct, Silas H. H. Clark, who thought it would be better to bring the price down gradually. In 1883 when telegraphers struck, demanding better wages, more reasonable hours, and equal pay for women—they were starved into capitulation. Gould merely commented that "the poorest part of your labor" was behind the effort. Later, when Gould reduced wages on his western railroads, the resultant strikes led to unprecedented bloodshed and damage. Gould gave little thought to the workers; to him the law of supply and demand dictated wages.[27]

Gould's social ethics were displayed again when gold was discovered in the Black Hills, which belonged by treaty to the Sioux Indians. He sent word to his associate Clark that if the find was significant, Clark was to build a road and stage line. The immediate result of the Indians' attempt to fight against the usurpation of more of their land was Custer's shocking defeat at the Little Big Horn River. Clark worried about the effect of the news on commerce. Dismissing the plight of the Native people, Gould wrote Clark urging him to push the project forward: "The ultimate result will be to annihilate the Indians & open up the Big Horn & Black Hills to development & settlement & in this way greatly benefit us."[28]

The leading New York newspapers—in particular Bennett's *Herald*; the *World,* after Gould sold it to Joseph Pulitzer in 1883; and the *Times*—assailed Gould at every opportunity. When Major J. R. Selover, a well-known stock operator who was said to be James Keene's man Friday, assaulted Gould on the street, the *Times* gave the incident front-page, detailed coverage. Selover, described as "huge and indignant," accused Gould of deceiving his friends. He and Keene had been involved with Gould in creating a bear market for Western Union stock. The belief on the street was that while they had been selling for a fall, Gould betrayed his friends by secretly buying. This drove the price up, costing Keene and Selover dearly. In accosting the tiny Gould, Selover rained blows on him while dangling him over a railing above a basement shop. When someone from the shop demanded that Selover stop, he allowed Gould to fall to the pavement below. Gould, said to be more frightened than hurt, soon avenged the incident in a transaction that caused Selover to lose $15,000 more. Of the beating the *Times* commented: "If reprisals like this are to be countenanced, Mr. Jay Gould will be hung by the nape of the neck and pummeled by indignant stock operators from January to December." Thereafter Gould traveled the streets with a bodyguard.[29]

One paper that did not attack Gould was the conservative and widely read *New York Tribune.* Gould had secured shares in it while loaning money for its purchase in 1873. Because he did not comment on Gould's Wall Street machinations, the *Tribune's* editor, Whitelaw Reid, who later served as ambassador to Great Britain, was referred to as Gould's "stool pigeon" or his "able lieutenant."[30]

Generally Gould's maneuverings in the news field were done behind the scenes. He publicized operations through rumors and planting stories. Often he offered stock tips to writers who slanted their views to support his interests or relayed information he wished disseminated. At other times he wrote anonymous letters or used pseudonyms to publish articles.

By 1883 Gould controlled the Union Pacific Railroad and Western Union Telegraph. He was forging a nationwide system of travel and communication. Western Union controlled telegraph traffic across the United States. Gould attempted to dominate the New York Associated Press by giving it special telegraphy rates. His proprietary alliance with the owners of the Atlantic cables gave him control of information from Europe. There were complaints that because his clique controlled the pool, there was no assurance of privacy when sending messages. One competitor was so certain Gould read all communiqués that he only sent ciphers, changing the code daily. In 1884 a Senate committee found that Western Union controlled commercial and financial news through a practical monopoly. Gould, regularly suspected of delaying the dissemination of information, was accused that same year of profiting by delaying announcement of the election returns. Because of his seemingly omnipotent power, some publishers saw Gould's interests as a threat to their very existence.[31]

Revisionist historians say that Gould was a businessman who was part of the fabric of the times, not the caricature of a robber baron that has been portrayed.[32] What is more relevant in this instance is how Mackay perceived him. Mackay would have been influenced by what associates reported—in particular, Bennett. Public perception would have played a part as well, and newspaper editorials and political cartoons about the financier were almost universally negative. Typical, as regarded telegraph and cable systems, was a February 1884 *New York Times* commentary that announced: "The desire to escape from the necessity of using lines even indirectly controlled by Mr. Gould has been growing strong in the United States."[33]

In proposing to build the Atlantic cable, Mackay attempted to de-emphasize the obvious challenge to Gould's monopoly. Mackay commented that it was not a strike against any existing company—it was purely an effort to give the public a cheap, reliable, and prompt cable service. When Mackay had approached other large cable users and capitalists to join in the effort, no one believed him and no one accepted. To potential investors the project was either a revolt against Gould or a patriotic sentiment. In either case it was not sound business. They did not want to risk their money against Gould, the man editor Joseph Pulitzer called "one of the most sinister figures that have ever flitted bat-like across the vision of the American people."[34] It became apparent the project could not be undertaken unless Mackay shouldered the burden. In the end he put in 70 percent of the money, Bennett about 25 percent, and acquaintances, including Comstock friend Harry Rosener and ex–mine operator George D. Roberts, the remainder.[35]

At the beginning of the financial war, Jay Gould's advantages were overwhelming. The monopolistic positions of the cable pool and Western Union appeared unassailable. Gould was the preeminent figure in the transportation and communications fields. His associates were men such as Pierpont Morgan, who sat on the Western Union board of directors. Gould had ex-

perienced managers in place, and his ingenious and merciless business conduct gave him the ability to set and spring traps at unexpected junctures.
Mackay had certain resources of his own. His personal wealth, used effectively, would allow him unusual power. It ensured investment funds rather
than leaving him at the mercy of the market or shareholders. His businesses
were largely unencumbered by the watered stock weighing down Gould's
enterprises. Mackay also began with the support of most of the public and
the press. His greatest liability was his lack of experience. He would have to
learn as he went and hope to survive the surprises of circumstance, as well as
those perpetrated by Gould. Adding to the instability of the land telegraph
situation was the existence of two other telegraphy enterprises: Bankers and
Merchants' and the Baltimore and Ohio. Their owners' actions in forming
alliances, setting prices, and expanding their lines would also play a large
part in Mackay's success or failure. In England a series of critical news articles questioned Mackay's role in the enterprise, observing that it was a
very large, hazardous undertaking for someone with no experience. Mackay
could only agree.[36]

Left, John William Mackay. Courtesy Nevada Historical Society; *right,* Louise Hungerford Mackay, a time after her second marriage. Courtesy Nevada Historical Society

The house Mackay built for Louise in Virginia City. Elegantly furnished, at the time it was worth twenty-five thousand dollars. Courtesy Special Collections, University of Nevada, Reno, Library.

James G. Fair, Mackay's partner. One of the most successful miners of the era, he was known on the Comstock as "Slippery Jimmy." Courtesy Special Collections, University of Nevada, Reno, Library.

Left, the astute James C. Flood; *right,* the amiable William Shonessy O'Brien. Barkeeps in San Francisco, they joined Mackay and Fair in the Bonanza Firm and became among the richest men in America. Courtesy Special Collections, University of Nevada, Reno, Library.

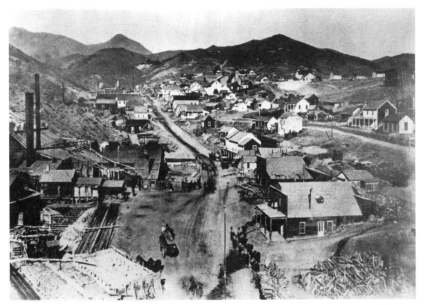

The French Mill, *in the foreground,* and the Sullivan Mill were purchased by Mackay with proceeds from the Hale & Norcross mine. Stockholders, supported by the *Gold Hill News,* which editorialized that the mine was "evidently paying somebody," called for an investigation into the mine's accounts. Collection 24,173; reproduced by permission of The Huntington Library, San Marino, California.

A view of Virginia City from the water flume built by Mackay and his partners. Collection 24,175; reproduced by permission of The Huntington Library, San Marino, California.

A Lawrence & Houseworth stereo-optic photograph of c Street, looking south from the International Hotel. Collection 713; reproduced by permission of The Huntington Library, San Marino, California.

The Con. Virginia's four-stack shaft house. The Con. Virginia and California works were rebuilt after the great fire of 1875, with the Virginia & Truckee railroad tracks circling them on three sides. Supplies were delivered on the uphill side, and ore was shipped out on the downhill. Courtesy Special Collections, University of Nevada, Reno, Library.

Miners at the Con. Virginia and California hoisting works. Foremen tended to be clannish in hiring American, Irish, or Cornish miners. Mackay instructed his foremen to treat every worker alike, saying, "All we want is good men." Courtesy Nevada Historical Society.

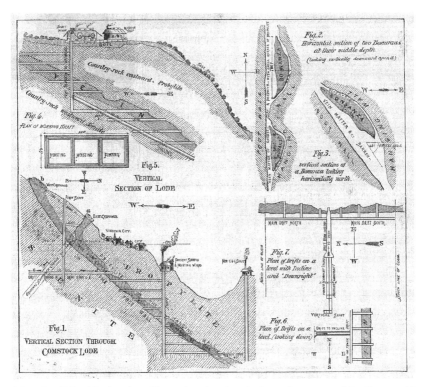

May 1877 *Scientific American* illustration showing the long, deep mass of the Comstock's "fissure vein." Courtesy Special Collections, University of Nevada, Reno, Library.

Leaving the Bonanza mines, October 28, 1879. Mackay, *far left*; Fair, *far right*; Mrs. U. S. Grant and General Grant, *in the middle*. Two years earlier, a reception for the former president at Louise Mackay's Paris residence helped launch Mrs. Mackay's social career. Courtesy Nevada Historical Society.

Louise Mackay was the first American to gain access to European aristocratic society. Although charitable in giving, she led a life of conspicuous consumption. One friend commented: "Mrs. Mackay never asked the price of anything." Courtesy Special Collections, University of Nevada, Reno, Library.

"SONGS OF OUR LAND."—Words and Music of "The Memory of the Dead."

M^CGEE'S ILLUSTRATED WEEKLY

DEVOTED TO

Art, Literature and General Information.
Entered at the Post Office, New York, N. Y., as second-class matter.

VOL. X.—No. 1.   NEW YORK, SATURDAY, MAY 21, 1881 $3.00 A YEAR. SINGLE COPIES 6 Cts.

EMINENT IRISH-AMERICANS.—J. W. MACKEY—THE BONANZA KING.

The Bonanza King at age fifty, featured on the cover of a news weekly. Courtesy Special Collections, University of Nevada, Reno, Library.

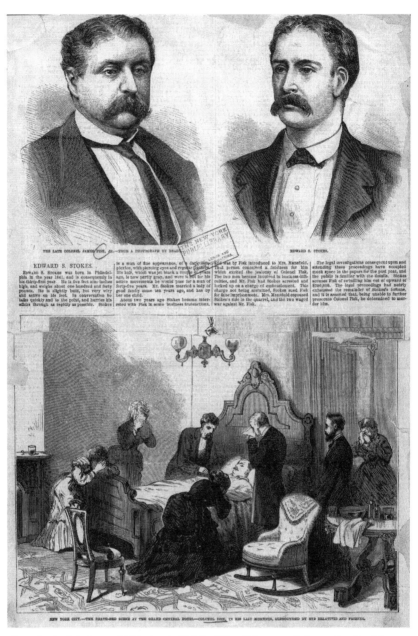

The principals in James Fisk's death. "Jubilee" Jim Fisk, *upper left,* was Jay Gould's closest associate. Edward Stokes, *upper right,* later joined Mackay in challenging Gould's telegraph cartel. Courtesy Print Collection, Miriam and Ira D. Wallach Division of Art, Prints and Photographs, The New York Public Library, Astor, Lenox, and Tilden Foundations.

Jay Gould's astute business sense and
ruthless ethics allowed him to monopolize
national and international communications. Courtesy Print Collection, Miriam
and Ira D. Wallach Division of Art,
Prints and Photographs, The New York
Public Library, Astor, Lenox, and Tilden
Foundations.

Uncle Sam confronts Gould, "the fellow that's to run all the
railroads and things." Government intervention seemed inevi-
table before Mackay challenged Gould's monopoly. Courtesy
Print Collection, Miriam and Ira D. Wallach Division of Art,
Prints and Photographs, The New York Public Library, Astor,
Lenox, and Tilden Foundations.

Breaking the transatlantic cable cartel. The second Mackay-Bennett cable is hauled
ashore at Manhattan Beach, near New York City. Courtesy Print Collection, Miriam and
Ira D. Wallach Division of Art, Prints and Photographs, The New York Public Library,
Astor, Lenox, and Tilden Foundations.

Clarence Mackay, who suc-
ceeded his father as head
of the Mackay international
communications network.
Courtesy Nevada Historical
Society.

John Mackay, nearing the end
of his life, continued to work.
The day he fell deathly ill,
he said, "I'll lay that Pacific
cable, and then retire from
business." Courtesy Special
Collections, University of
Nevada, Reno, Library.

The statue of John William Mackay, by renowned sculptor Gutzon Borglum, stands before the Mackay School of Mines on the University of Nevada, Reno, campus. Courtesy Special Collections, University of Nevada, Reno, Library.

CHAPTER 12

The New War

MACKAY BEGAN his attack on Gould's system by investing heavily in an overland telegraph company. In researching the Atlantic cable he came to realize that the opposition would control how quickly and accurately a message coming through his lines would be transmitted once it was in America. No matter how efficient the oceanic transmissions, no one would use the service if messages sat undelivered on land.

Western Union, formed in 1866 by the merger of the three major telegraphy enterprises, was the first nationwide multiunit business in the United States. It controlled a large percentage of the lines throughout the country, and none of its opponents had the resources to compete effectively. In 1881 Norvin Green, Western Union president, told his stockholders that the company had gained such magnitude and strength that competition could "not materially interfere with remunerative dividends." In August 1883 Mackay formed a syndicate and purchased 12,000 to 21,000 shares in the failing

Postal Telegraph Company. He had been urged to do so by limited cable partner George Roberts, who was already a Postal Telegraph investor.[1]

The following month Mackay went forward with the plan to build a transatlantic cable. He commissioned the England cable makers Siemens and Brothers (who employed 2,500 workers and whose buildings covered 7½ acres) to build two submarine cables totaling well over 6,000 nautical miles. At the same time he contracted with a company in Scotland to build a cable-repair ship, the CS *Mackay-Bennett*, to be based at Halifax, Nova Scotia.[2]

His land-based operation, reorganized on October 19, 1883, had originally been named Postal Telegraph in order that patrons might associate it with a government proposal for a U.S. Post Office telegraph system. As far back as 1872, in his annual report, the U.S. Postmaster had recommended that a telegraph system be developed under his department's jurisdiction. In January 1873, in the Far West newspapers from Petaluma on the West Coast to Carson City, Nevada, clamored for government intervention. They chided Western Union and their "lackeys at Associated Press" for what was termed an unholy alliance that dictates who is able to present the news.

The *San Francisco Alta* assailed the telegraph service, charging that as soon as they spoke out for telegraph reform, Western Union became "determined to deprive us of the privilege of receiving dispatches." At the time the *San Francisco Chronicle* and the *Alta* shared items from the East, because sending one long article tied up lines until late at night. When Western Union charged the companies an extra 50 cents per word for sharing items, they refused to pay. Western Union's response was to refuse them service. While still contending that the fee was an overcharge, the *Alta* attempted to pay, but the Western Union supervisor would not allow it. He said the matter had been referred to the home office and he could do nothing more. The *Sacramento Record* reported that the telegraph monopoly had commenced open war upon papers that advocated a U.S. postal telegraph. The *San Francisco Evening Post* related how the *San Francisco Herald* was crushed from existence by order of Western Union and how the *Sacramento Union* was so intimidated by it in the early 1860s, "the *Union* will not say a word to this day."[3]

In 1881 James Keene thought to break Western Union's death grip on communications. When he arrived in New York, he brought millions of dollars—much of it coming from acting the bear against Flood and Mackay's bonanza stocks. In 1880, when Keene attempted to corner the wheat market,

Gould had destroyed him. The idea of cornering the wheat market was to trap bear speculators who sold grain for future delivery. An operator might corner the market by buying all the grain in the market so that bear speculators had to buy from him at exorbitant prices when their delivery was due. Before Keene could sell his wheat, Gould drove the market down, and Keene lost $7,500,000. Legend has it that Gould commented: "Keene came from California in a private car. I will send him back in a box car." Keene, who was also becoming a world-class breeder of racehorses, could be beaten but never defeated. In July the following year he reappeared as president of the Postal Telegraph Company.[4]

One of Keene's associates in the telegraph company was George D. Roberts, who had been his partner in creating the bear market that damaged the Bonanza stocks. Roberts had been among the earliest arrivals in Virginia City, riding there from Nevada City in July 1859. He and Mackay knew each other from that time. In July 1881 Keene, Roberts, and their associates were intent on revolutionizing telegraphy, stating in their prospectus that within two years they would operate a system the equivalent of Western Union. They paid $1,000,000 for the patent of W. A. Leggo for his "Electro-Graphic automatic and fac simile telegraph." The machine allowed drawings, music, and signatures to be reproduced across the wires. They also purchased Elisha Gray's harmonic duplex system. This invention used tones of different pitch that allowed six messages to be sent simultaneously over a single wire. Gray's system was first offered to Western Union but was purchased by Postal for $1,500,000 when Western Union haggled over the price. Gray's invention led to the belief that the business of the country would soon be conducted over 30,000 miles of wire instead of the current 250,000 miles.[5]

Unfortunately for Keene, he was a better investor than manager. For two years Postal Telegraph struggled with a piecemeal system, stymied in its attempts to implement full operations. Keene was on his way to another overwhelming failure—which would be followed by another rebound in the stock market in the early 1890s. By July 1883 Keene was ousted, and ex–U.S. Marshall Joel B. Erhardt assumed the telegraph company's presidency.

The company's line between New York and Chicago remained unfinished. The company lacked organization. It owned the valuable patents, but as the *Times* reported, "a want of system at head-quarters had rendered the outlook foggy and confused." Erhardt was collating facts to report on how it might be reorganized. His job was complicated by a strike by the fledgling telegraph operators union against Western Union and the other large telegrapher, the Baltimore and Ohio Telegraph Company. The workers' grievances included unequal pay for women, low wages, and no recompense for overtime or working Sundays. One Western Union official stated that rather than submit, "the Company will close every office." Postal's New York–to–Chicago line was hurriedly strung together, and urgent messages and business messages were sent across it free of charge, allowing communication between Wall Street and the Chicago Board of Trade to continue. Postal's board was arguing over charging the same rates as Western Union once Postal was in full operation or offering cut-rate fees. The issue was resolved on August 15, the same day the telegraphers' union strike failed and members returned to work. The board voted to charge lesser rates, and Erhardt, the company secretary, and two directors resigned. Mackay took charge, and rates were cut.[6]

Roberts had begun recruiting Mackay in the fall of 1882, urging Mackay, who was on his way to Europe, to investigate the company. Mackay arranged for two experts to examine its affairs while he was away. For over a year, he and Bennett had been contemplating laying the Atlantic cable. Apparently finalizing those plans, he returned and completed his analysis of Postal Telegraph. Mackay declined comment to the press, but voluble market sharp Henry Clews observed: "If Mr. Mackey [*sic*] goes into the Directorate of the company he means to back it up with his wealth. In that case the company will be a very formidable competitor of the Western Union. . . . It is a blow struck at a monopoly."[7]

Mackay, Roberts, and prominent and powerful bankers H. L. Horton, George F. Baker, and George W. Coe formed a syndicate and became majority shareholders of Postal Telegraph's capital stock. The shares were placed in a trust to be untouched for three years, with Mackay as custodian. The syndicate would furnish whatever money the company needed. Mackay understood a basic tenet of corporate business—that most owners did not have the experience, time, or information of full-time career managers. He certainly did not. Following the same course as in his mining and banking endeavors, he refused the presidency of the operation. But within two weeks, on August 29, 1883, being looked to as the controlling individual, he reconsidered and assumed the office. The next day $150,000 in contracts were awarded for the construction of new telegraph lines.[8]

Russell Sage, the shrewdest of stock traders and Gould's trusted advisor, was reclusive, crafty, and another large investor in Western Union. He had known Mackay and Roberts during the Comstock days. In his judgment their new company would fail. "This new blood won't amount to anything," he said. He remarked that he hoped the stockholders would fare better than he had when he was involved with Mackay in the Consolidated Virginia and California mines: "The great trouble with us stockholders was that we could not see so far into the rock as Mackay could. So we got left, and Mackay got rich." As for the new corporation's being able to compete with Western Union, he thought back to previous rivals: "All of those have come to us . . . and have been anxious to sell out." At the end of the interview, Sage smiled and added, "Put that in the paper about Mackay and his mining business, sure."[9]

In 1883 and 1884 Postal Telegraph was putting up poles and stringing wire, but they were limited because Western Union had exclusive rights along almost all railroad lines. It had gained private usage in exchange for allowing railroads free use of its wires to coordinate schedules. The free use was so valuable that railroads also allowed the operation of Western Union telegraph offices in their depots. By 1870 nearly 9,000 of Western Union's 12,600 offices were located in train depots. As the giant telegraph concern now blocked Postal Telegraph's extensions, Postal was compelled to string lines along roadways and across private properties, requiring permissions and buyout agreements with local governments and innumerable land-owners.[10]

The telegraph business was cutthroat. Competition included not only allowing franking privileges for members of Congress, but also giving rebates to businesses for use of a particular company's lines. While Western Union issued complimentary franks to national, state, and municipal authorities, Postal Telegraph for years restricted franks to certain members of Congress but later extended the privilege to judges, aldermen, and mayors. The largest rebates were given to the most demanding companies and continued until several years later, when a truce in the telegraph wars was declared and uniform rates were established.[11]

Since December 1882 James Bennett's *Herald* had been attacking Gould's telegraph and cable pools as "the most gigantic system of organized robbery in existence." Throughout February in 1884, the paper devoted space almost daily to attacking the monopolies. At the same time Congress was nearing a decision regarding whether it should break up Western Union.

Senator George F. Edmunds of Vermont introduced a bill for the erection of government telegraph lines. Edmunds, a member of the Senate Judiciary Committee, had earlier sought legal action against Gould and the other directors of Union Pacific for paying unauthorized dividends. He now argued that in promoting the general welfare, Congress needed to provide for the dissemination of intelligence—that instantaneous transmission of information should not be subject to censorship or corporate will but, like postal service, should be fair to all. He pointed out that the government could not be squeezed or bought out, as was the fate of companies that had attempted to compete with Western Union.[12]

Also in mid-February, Mackay hired George G. Ward as general manager of the Atlantic cable company, incorporated as Commercial Cable Company. Ironically in a system touted as "all American," Ward was an Englishman. He was also an experienced and capable administrator who would serve as manager and vice president for over twenty-five years. Ward had been working as general superintendent in America for the Direct United States and the French Atlantic Cable Company—part of Gould's cable pool.[13] The new job would tax Ward immediately—the war between Jay Gould and Mackay was commencing.

In 1884, besides Mackay's reorganized Postal Telegraph, there were two challengers to the Western Union land telegraph system: A. W. Dimrock's Bankers and Merchants' Telegraph Company and the Garrett family's Baltimore and Ohio Telegraph Company, which was tied to their successful Baltimore & Ohio Railroad. Both telegraph companies were attempting to expand. Bankers and Merchants' acquired a fourth leading company, American Rapid Telegraph, in August 1883, and each company began stringing lines on the other's poles. Early in 1884 the Baltimore and Ohio purchased two lines in New York State.[14]

In May 1884 a stock panic injured the challengers' ability to oppose Western Union. Three banks and at least eight brokerage houses failed (another bank had its president flee to Canada after losing $4 million of the bank's money in the crash). The stock of Dimrock and Company, parent of Bankers and Merchants' Telegraph, dropped from 119 per share to 45, dragging the telegraph company down with it.

There were rumors that Western Union would buy out the failing entity; instead Mackay stepped forward, bringing together Dimrock with Robert Garrett, of the Baltimore and Ohio, and himself. For several weeks they met in negotiations. Finally on July 17, 1884, they formed an alliance consolidating the three telegraph companies into one great system. Together the companies would control 120,000 miles of wire, connecting all the principal cities east of the Missouri River and southward to Galveston, Texas. Although less than half the length of Western Union lines, because they reached all the important trade centers, about 80 percent of the country's paying telegraphic business could be covered. The three companies, to be called the United Telegraph Lines, would be run as nearly as possible as one, using Postal's rates—which were significantly lower than those of Western Union.[15]

The month of July seemed to augur nothing but good for Mackay. On the 19th the first cable of his Commercial Cable Company was complete from Ireland to the United States. Work on a second cable was on schedule; it would be laid in two and a half months.[16]

At the end of August the U.S. Postmaster-General met with representatives of the new United Telegraph as well as those of Western Union. United had proposed a 100 percent reduction in the land telegraph rate charged the government. The Western Union representative protested, saying his company already provided the government with the lowest rates offered to the public. This statement was disproved with figures showing that the government, paying $500,000 a year at rates fixed by the company, closely approached the highest rate charged to the public. It was pointed out that when competitive lines were built Western Union reduced rates, while in uncontested areas high prices were maintained. In commenting on the meeting, the *New York Times* remarked that the United Telegraph proposal mirrored rates suggested as fair by the U.S. Senate. The competition, the *Times* suggested, would go a long way toward "practically solving the question of low rates without governmental enactment."[17]

At the time of the meeting, Mackay had invested approximately $5 million in his communication system. In places where they were in direct competition with Western Union, rates for newspapers using telegraphy had been driven from a penny a word down to a quarter of a penny. Roberts told a newsman in Chicago that Postal Company's great mission was to cheapen and popularize the wire until it would almost entirely supplant Post Office business.[18]

Early in September 1884, a mere two weeks after the meeting with the Postmaster-General, Mackay's United Telegraph partnership unraveled. Owing to the market crash, Dimrock's Bankers and Merchants' could not follow through on the agreement, and Robert Garrett pulled the Baltimore and Ohio Company out. Mackay attempted to sustain the partnership with Dimrock's company, but three months later Bankers and Merchants' internal problems caused the dissolution of that agreement as well.[19]

Two weeks after Robert Garrett's action, his father died. John Work Garrett had been the force behind the Baltimore and Ohio success. Robert Garrett was a Princeton alum, well known in social circles and owning various expensive residences—including one in Baltimore said to be the most expensive ever built in that city. Courteous and affable, he did not have his father's aptitude for business. Despite withdrawing from the telegraph pool, he continued stringing lines and building offices for Baltimore and Ohio Telegraph.[20]

Mackay's troubles multiplied on October 7, 1884, two days before his second Atlantic cable was completed: word came that his first cable had broken. Two days earlier it had been announced that two of Gould's American cables had broken. The cause of the Gould cables' breaks was unknown. Icebergs were suspected in the Mackay case. The broken Mackay cable was repaired two months later, while Gould simply used other cables in his pool, his broken cables remaining in disuse until the following July. It was later asserted that the breaks were the result of the second Mackay-Bennett cable's being laid across the others. (While picking up Gould's cable for repair, at the end of July 1885, the second of Mackay's cables was broken. The new repair steamer, the *Mackay-Bennett,* was dispatched, and the broken cable was quickly restored.)[21]

Mackay's competitors were creating other problems that would haunt him. Robert Garrett had begun investing in developments that his deceased father had found too costly and risky: establishment of an express company to compete with the well-established Adams Express; building and running parlor cars in competition with the Pullman Company; and obtaining an entrance for his railway into New York City in opposition to four of the most important railway systems in the country. Jay Gould was closely watching Garrett, poised to strike.[22] Gould was interested as well in the actions of Mackay and a new player who had joined him: the man who had slain Gould's partner Jim Fisk—Edward Stokes.

On February 1, 1885, after Mackay's efforts to pool telegraph resources against Western Union had failed, Postal Telegraph defaulted on its bonds. Mackay was disposed to close out the entire land-telegraph business but did not follow his instincts. His managers, Albert B. Chandler and Henry Rosener, were appointed receivers. He had, a time earlier, helped Stokes (his friend from San Francisco) purchase the Hoffman House, a successful hotel and restaurant in New York City. Stokes was still handsome and confident, but he was haunted, apparently fearing assassination at the hands of Jim Fisk's friends. In public he always ate with his back against a wall.

Mackay confided to Stokes that he had sunk over $1,500,000 in Postal Telegraph and was afraid he would take a heavy loss. Stokes suggested that instead of cutting his losses, Mackay could make the company pay. Stokes knew the insolvent Bankers and Merchants' Telegraph Company, also now in receivership, was an extensive and valuable property. If Mackay advanced him money, he would buy the Bankers and Merchants' receivers' certificates with a view to acquiring its properties to combine with Postal Telegraph. The plan mirrored Mackay's original proposal, and buoyed by the thought, Mackay rescued Postal by refinancing it. At the same time he advanced Stokes $100,000, half as a loan to the receiver on the security of a first lien on the Bankers and Merchants' wires.[23]

On the Comstock Mackay had utilized Fair's evaluations underground and deferred to Flood's financial expertise. In his communication businesses he would rely on managers for day-to-day decisions but kept his own counsel in top-level decision making. An unnamed associate once said Mackay was more difficult to sound out than any man he had ever known, adding that "beyond a certain depth he remains inscrutable." Mackay's office walls were covered with photographs of distinguished Americans he counted as friends, but there is no indication that he ever asked advice or assistance from any of them. Likewise, among his few intimates, none ever mentioned his using their counsel. Mackay was intelligent and confident. He planned with care and was devoted to detail. When approached by visitors, he was attentive but brief and decisive, making split-second decisions. And he asked no quarter from his opponents.[24] But in this instance his equanimity was shaken.

Shortly after he fronted Stokes the money for Bankers and Merchants' Telegraph Company, he again realized how suffocating the competition from Western Union was. Because of their exclusive rights of way and lines established across the country, their dominance was unchallengeable. He wrote Stokes, saying he would rather lose a couple of hundred thousand dollars than see Stokes "stuck," but there was no way to win. He concluded: "You are young in the business. Like myself know nothing about it and a man that goes into business he knows nothing about always fails in it." Mackay said it infuriated him to think how "thieves and confidence men like Roberts" had duped him into becoming involved. "I ought to have been tied to the tail of a cart and whipped like a dog for being roped into it." But Stokes insisted that victory was in sight and, saying he was master of the situation, asked for further advances to secure consolidation plans. Mackay unenthusiastically acceded.[25]

In May 1885 the Bankers and Merchants' receivership, buoyed by the Stokes-Mackay investment, was under the supervision of a competent manager, General J. G. Farnsworth. A separate receiver, Edward Harland, was running American Rapid (the merged company that shared Bankers and Merchants' poles and lines). Harland thought he might interest Western Union in American Rapid. Because of the shared poles and lines, Harland asked the courts to determine which properties belonged to which company. State Supreme Court Judge Charles Donohue, who had a link to the Tammany Hall scandal—ten years earlier he had freed William M. Tweed from incarceration by delaying proceedings against him—agreed to separate the lines. The process would take time, and Donohue gave his assurance that Bankers and Merchants would be allowed to protect their interests before American Rapid was permitted to enter into a contract with another company.

Mackay's other interests left him short of cash: over the course of the previous months he had spent $4 million in buying Fair and Flood's shares of bank stock, as well as millions in the new cables submerged in the Atlantic and large sums on the Comstock in deep mining and the reworking of the established mines. On January 1, 1885, Robert Garrett introduced sweeping rate reductions in the Baltimore and Ohio telegraph rates. The ruinously low charges injured the Mackay interests more than Mackay's competition with Western Union. Mackay saw that Garrett was playing a losing game. "Gould hates Garrett bad," he observed, "and will make no terms with him." Toward the end of May 1885, he advised Stokes to try to sell the Bankers and Merchants' plant to Western Union "ticker and all." Saying he did not care about the money already invested, he continued, "I want to close it out and have no more bother with it. If I can't do this, I will give it away." Stokes wanted to persist and requested more funds. On May 31 Mackay wrote that it was one thing for Stokes to plan and scheme but another to raise the money to carry the plans out. He complained: "I have done nothing but pay out coin the last four years. I am getting short now." But the letter also allowed Stokes hope. Instead of an outright refusal, Mackay replied only that he could not advance money for thirty or sixty days.[26]

Gould had been biding his time. The Stokes plan for Bankers and Merchants' was to reorganize it on a basis that would keep American Rapid under its control. Bankers and Merchants' stockholders had agreed to sell the company on June 10 to Stokes and Mackay's new United Lines Telegraph Company (not to be confused with United Telegraph Lines Company, as the previous consolidation was called). Stokes declared that the corporation would begin with a capital of $3,000,000. The other holders of receivers' certificates and the company's bondholders agreed to support the sale. But on June 9 an investor described by the *New York Times* as "the holder of a paltry sum of bonds" requested a hearing on the sale. Judge Donohue ordered the sale delayed.

In the meantime Gould summoned American Rapid's receiver Harland to a meeting on his yacht. The men agreed to a contract stipulating that Western Union would act as American Rapid's agent and take control of its lines. Gould's attorney, with the contract and a list of what American Rapid claimed was its property, approached Judge Donohue. The judge, breaching his guarantee that Bankers and Merchants' would be given the opportunity to protect their interests, issued an order that the Bankers and Merchants' receiver, Farnsworth, had to deliver said property to Western Union.[27]

On July 10, the day the company was to have been sold to Stokes and Mackay, Gould's agents confronted Farnsworth at the Bankers and Merchants' offices on Broadway in New York City, presenting him with a court order declaring all American Rapid lines were to be turned over to them. Since the judicial process of separating the wires had not been done, Farnsworth would not comply. A time later Farnsworth left the premises to consult with counsel, and Gould's men—variously described as a platoon, a posse, and a force—reentered and cut nearly all the lines in all directions. The severed cables were dragged across the street and run into the Western Union building. In Albany, Rochester, Troy, and Baltimore, Maryland lines were similarly cut and taken onto Western Union property. While the raids were occurring, the counsel for Bankers and Merchants' attempted to obtain a restraining order. Judge Donohue could not be found.

That night in Albany agents from Bankers and Merchants' recaptured lines there, and they were attempting to do the same in Rochester and Maryland. With things escalating out of control, Donohue turned up and issued a restraining order commanding that neither side interfere further with the other.

When the Bankers and Merchants' lawyer issued the restraining order against Norvin Green, president of Western Union, Green—who twenty years earlier had been one of three men to lead the merger of telegraph companies to create Western Union—claimed not to know anything about the incident. The lawyer pointed out that Green had signed the petition on which the order was granted. Green protested that he had signed merely because attorneys had told him to do so. Finally Green agreed to see that notice of the injunction was communicated to Western Union operatives. But his notification was not sent until the next morning, by which time more wires had been confiscated and the Albany lines retaken by Western Union agents.[28]

Mackay ordered that Postal Telegraph lines be made available, as did Garrett with the Baltimore and Ohio, and Bankers and Merchants' was able to carry on most of its regular business. The *Times* described the Bankers and Merchants' Broadway office: "The instruments clicked busily all day right under the ears of the omnivorous rival just across the street." Obviously the object of the raid was to cripple Bankers and Merchants' Telegraph Company. In a statement by a press agent issued two days after the raid, Western Union—while not admitting any of the wires cut belonged to Bankers and Merchants'—allowed that if proved otherwise, "that company has its redress in the courts." Attempting to de-emphasize the scope of the atrocity, the agent stated further that for some time Bankers and Merchants' had been negotiating to sell its lines to Western Union. Dwight Townsend, the chair of the Bankers and Merchants' reorganizing committee, was emphatic in his refutation, saying simply: "It is a lie."[29]

President Green telegraphed company superintendents, explaining that they had taken the lines under court order and that until a further order was received, property in possession of Western Union should not be returned. Townsend described the seizure as an "iniquitous act on the part of a great monopoly." He went on to say that though he had been charged with saying that Judge Donohue had high regard for Jay Gould and Western Union, "I have not made that statement publicly, whatever my opinion may be on the matter." The following spring, charges of corruption over other actions taken by Judge Donohue were referred to the legislature, and the *New York Times* reported that he was only saved because his friends in the assembly produced a "whitewashing report." In 1887 he lost his bid for re-election to the court. As for Edward Harland, the receiver who had set the episode in motion by signing over American Rapid's assets to Gould, he had disappeared.[30]

Once a degree of normalcy was restored and certain of the lines were rebuilt, Stokes went forward with his plan. At the court sale of Bankers and Merchants' he used receivers' certificates as partial payment and bought the controlling interest. He became president of the new corporation: United Lines Telegraph Company.[31]

Western Union continued to harass Stokes and his new company. When Stokes challenged the company's Wall Street ticker service by offering brokers and bankers a more efficient ticker system, Western Union officials attempted to have the stock exchange refuse the new company exchange quotations. They also published circulars telling customers the new ticker service was a patent infringement. Lawsuits followed, each won by United Lines. When Stokes ran wires to connect brokers' offices at Hoffman House with the new tickers, Western Union again attacked. On Christmas morning 1885, six miles of recently strung United Lines wire was cut and removed. Stokes immediately got crews working to replace the lines and filed affidavits to apply for arrest warrants. The attack had been perpetrated on the Christmas holiday so the courts could not interfere. Calling it "high-handed robbery" and saying that he was being hounded, Stokes commented further that it reminded him of the days when Fisk, Gould, and Boss Tweed's corrupt ring controlled the courts. "They don't seem to respect God, Judge or devil," he said. The superintendent who ordered the wire cutting responded that they were foreign lines on Western Union poles. Two days later, the *New York Times* reported that Western Union had made a mistake and was "kindly informing the United Lines Company that it can send for its wires."[32]

Three weeks earlier, on December 8, 1885, an occurrence altered Baltimore and Ohio president Robert Garrett's fate, setting in motion events that would insure the ultimate triumph of Western Union in the telegraph war. Garrett, in difficult financial straits, was meeting with William H. Vanderbilt in the hopes of securing Vanderbilt's assistance in gaining a New York terminal for the Baltimore and Ohio Railroad. While assuring Garrett that he would assist in the project, Vanderbilt was suddenly stricken and plunged headlong to the floor, dying of apoplexy. The shock of the scene, along with Garrett's dashed hopes, seemed to unbalance the troubled mogul.[33]

Several months later Mackay approached Garrett, reintroducing his plan to include the Baltimore and Ohio Telegraph in a merger with Postal Telegraph and Stokes' new United Lines. After interminable, intense negotiations Garrett agreed to terms, but before signing the contract, health problems caused him to sail for Europe. When he returned, he had changed his mind. Citing only vague reasons, he refused to join the consolidation. Later, after his company had been sold out from under him, he complained that his associates in the Baltimore and Ohio syndicate wanted a better deal by 5 percent and would not compromise. Mackay and Stokes were left dumbfounded.[34]

Western Union was suffering from Garrett's erratic decision making as well. Its attempts to compete with his radical reductions in fees, as well as the added competition of the Postal lines, cut deeply into its profits. From 1883, when Western Union brought in $7,500,000, earnings fell to under $4,000,000 in 1886. Consumers benefited, as average tolls over the period fell from 38¢ to 31.[35]

In the meantime Stokes and receiver Farnsworth had filed suit against Western Union, demanding $2,000,000 in damages for the attack on the Bankers and Merchants' wires and cables. Western Union took its plea for a stay to three different courts, failing in each attempt. On May 2, 1886, the *New York Herald* reported that each of the judges found the plea was nothing more than a pretext for delay. "The lawyers of the monopoly finally trooped before Judge Wallace, of the United States Court, but their nervous appeal was yesterday summarily dismissed by him also." On May 7 Mackay wrote Stokes: "You are in now for a good fight with the Western Union, and you must be prepared to win your suit. You have a tough gang to fight and you must up and at it."

On May 26, on a map on the wall in New York's Supreme Court, circles were drawn representing all major cities east of Kansas City. Red and black lines extended from the circles, delineating the telegraph cables of Bankers and Merchants' and the American Rapid Telegraph Company. The diagram was described to spectators, who sat two to a seat with at least a hundred standing during the proceedings. Eight renowned attorneys represented the two sides. At the Bankers and Merchants' table was powerful ex-senator Roscoe Conkling, ex-boss of the government spoils system, while counsel for

Western Union included two ex-judges and Joseph H. Choate, renowned as the era's "idol of all court lawyers." Officers of the Western Union sat at the defense table taking notes of the testimony.[36]

The trial lasted eight weeks. Plaintiffs' attorneys described how the telegraph wires were cut indiscriminately, estimating that $6,150,000 in damage was done to Bankers and Merchants' Telegraph Company. The defense produced the order from Judge Donohue instructing Receiver Farnsworth to deliver the property of American Rapid Company and presented myriad papers—1,500 were eventually introduced—dealing with the weighty subjects of telegraphy and cable ownership. In closing arguments Choate compared the jurors to Job, thanking them for their patience and suffering. He then tested the analogy by spending several hours explaining his theories of the case. In the end the jury found for the plaintiffs, awarding $240,000 to Stokes and the Bankers and Merchants' Company. Choate immediately asked for and was granted a stay of 120 days.[37]

The finding notwithstanding, Gould and Western Union had won. Bankers and Merchants' service had been severely hampered for ten days and was not restored to full service for over a month. The disruption had cost not only a considerable amount of money, but also the loss of contracts. Mackay's plan to pool resources with Bankers and Merchants' and Garrett's company had been so severely retarded that it would never be implemented. Mackay wrote Stokes complimenting him for a good fight. "Wait and rest and think awhile and see the next best move," he counseled. "In a year from now they will have a belly full of litigation and cheap rates."

Gould saw things differently. July 11, 1886, the day after the verdict, he wrote his son: "The jury in the WU $2,000,000 suit after being out all night rendered a verdict of $240,000 which we will of course appeal. . . . We can carry it to the U.S. Court at Washington which will take 7 years before Mr. Stokes will see any cash."[38]

CHAPTER 13

Near Disaster

WHEN THE SECOND Mackay-Bennett cable was to be landed at Manhattan Beach, just east of the Oriental Hotel, in mid-October 1884, 850 people braved a cold, bleak wind to wait on the shore. Special trains were run carrying scientists and electricians from around the country, invited to watch the completion of the first cable to reach a point so near New York City. A big red flag, with the initials C.C.C. for Commercial Cable Company, floated atop the sending and receiving cottage, built on pilings 100 yards from the water. The steamer *Faraday* had left Waterville, Ireland, a month earlier, laying cable as it made its way across the Atlantic. (From Waterville the company had branch offices in London, Liverpool, Birmingham, Glasgow, and Havre.)

The spectators were disappointed—the *Faraday* dropped anchor two miles from shore. A tug that was to tow crafts carrying the coiled cable into the beach met the ship. But stiff winds and a short, choppy sea caused the guide rope to get caught in the tug's screw, and the afternoon was spent attempting to untangle it. George Ward nevertheless went ahead with the ceremony, giving a speech calling attention to the fact that Commercial Cable was going to fight monopoly and bring cheap rates to cable transmissions.[1]

It was two days before the cable was landed. Heavy breakers drove the boats off course and upset them in the icy water, delaying the process. Finally, on October 18, the tug ran the cable to within 150 yards of the breakers. It anchored, and two small boats—between which was tied a rubber raft carrying the heavy wire—started for shore. A full compliment of oarsmen in the boats rowed, while hands on shore pulled at a rope attached to the raft for that purpose. As it cut through the roiling breakers, a gang of thirty men waded out to their waists in water to haul the cable ashore. Hoisting it on their shoulders, they dragged it through a trench dug in the sand up to the small office while hundreds of spectators cheered.

The *Faraday* steamed away to splice the shore cable to the main line. It then proceeded to Nova Scotia and out to sea to find and repair the broken portion of the first Mackay cable. On December 6 the splicing of the break was completed, and that afternoon the executive office of the Commercial Cable Company in America was in communication with Ireland.[2]

Before his cables were in operation, the management of the pool cable lines had continually tried to resolve the issue of fees with Mackay. They insisted his company must adopt the 75¢-a-word rate the rest of the lines charged. If it did, it could have the share of business it might win; if it did not, there would be reprisals. Mackay refused, and the pool companies dropped rates to 50¢ a word. On Christmas Eve 1884 the lights were on at 21 Wall Street, the offices of the Commercial Cable Company, better known as the Mackay-Bennett Company. Its cable service from Europe to America had begun. Mackay charged 40¢ a word.[3]

The next day the pool companies announced a reduction to meet Mackay's charges, but businesses and newspapers on both sides of the Atlantic gave the new service a trial. The *Times* announced: "The old combination has undoubtedly been aroused from the lethargy which has for the past two years possessed it." But the paper warned that the Mackay-Bennett people would need to display extraordinary strength to maintain their independence. There were rumors that negotiations seeking to add Mackay's company to the fold had already begun.[4]

For a year the Commercial Cable Company was successful. At 40 cents a word they were charging what they believed the service was fairly worth.

In doing so, they earned over $1,000,000 in cable business. The pool of competing companies—Anglo-American, Direct United States, French Atlantic, and Western Union—repeatedly approached Mackay about joining them, but Mackay vowed never to do so. The public lent moral support. In September 1885 the *Electrical Review* commented: "John W. Mackay and James Gordon Bennett are said to have done more than any other two men to lessen the evil influence of Jay Gould. The Commercial Cable is the sharpest thorn in Jay Gould's side. Rates have been much reduced and the service greatly improved."[5]

The pool cable companies, having to pay large interest on watered stock, could not pay dividends while charging Mackay's rates, and overland telegraph profits had shriveled as well. Gould was undeterred by either circumstance; he rallied his associates and intensified the conflict. In mid-April 1886 the pool companies met secretly. In May they began an offensive they believed would force Mackay to capitulate or face ruin. Their goal was to bully Mackay either into the pool or into a selling his properties. They cut cable rates to a ruinous 12¢ a word—below the cost of transmission. Western Union made the squeeze possible by foregoing earnings, thereby insuring their own deficit for the year.[6]

Except for a ten-day trip to Paris (presumably made to see Bennett), Mackay was with his family in England in the spring of 1886. He was with Bennett when word of the cable monopoly's action reached them. Their first thought was to answer the move by reducing their charges to 6¢ a word, but upon further consideration they made a bold and brilliant decision.[7]

On April 30, 1886, in New York, Commercial Cable general manager George Ward issued an announcement on rates. He began by relating how the company had repeatedly resisted attempts by the other cable services to coerce it into raising its rate. The pool's new rate of 12¢ a word insured losses that could only be regained by largely increased tariffs once the rate war ended. Reasoning that their customers were those in the business community who knew competition was necessary to keep long-term rates low, Ward declared that Commercial Cable would not meet the competition's rate. Instead it would reduce its rate to 25¢ a word. Its patrons must decide whether it was in their best interest to utilize the monopoly's lines at the temporary cheap price, knowing what would happen to rates if Commercial Cable was destroyed. The alternative was to pay the Commercial Cable rate and ensure continued competition, from which all had benefited.[8]

Bennett campaigned vigorously in the *Herald* against the monopoly's attack, noting that the cut to 12¢ was a desperate attempt to force Commercial Cable into the pool so tolls could again be raised. He called the war one of "monopoly and extortion against independence and fair prices." The *Times* of May 2, 1886, announced that a large number of prominent bankers, brokers, and merchants had declared their intention of continuing their patronage of Commercial Cable regardless of the reduced price offered by the monopolists.

The *Herald* printed a long list of testimonials from those supporting Mackay's company. "Hurrah for the Mackay-Bennett cable," said a message from an importer of chemicals, "[it] will continue to get all our business at twenty-five cents a word, or fifty cents, for that matter; for we believe it offers us the only safeguard against extortion." Another company executive commented: "[Commercial Cable] deserves support by every right of business, sentiment and good morals. It is resisting an assault that is transparently for the purpose of making it possible to establish extortionate rates in the future." A banker declared: "Believing that Mr. Mackay and associates intend to maintain independent relations with the public at reasonable rates, and that the reduction of the pool cables is for the purpose of forcing a combination, we shall continue to patronize [Commercial Cable] irrespective of the temporary rates by other lines." A dozen other companies were quoted, including one owned by the mayor of New York and his bother.[9]

On May 13, 1886, the Board of Trade and Transportation passed a resolution saying the public should patronize the Commercial Cable Company, because through it lower rates had been attained. On the 15th economic pundit and regular contributor to the *Times* J. S. Moore remarked in the paper that the pooling of the other companies had come about after a similar fight some years earlier. The reward to the projectors and manipulators, he observed, came in the form of diluting stock, bonds, and preferred shares, which were the rewards of pooling and mergers. Moore went on to say that the Mackay-Bennett cable was not open to combination. It was a cable laid for business, not merely to be sold like certain of Gould's rail lines. He concluded that the 12¢ rate was a ploy resembling "the farmer's wife's invitation to the ducks, when she throws them some crumbs with the sweet ejaculation, dilly, dilly, dilly, come here to be killed."[10]

In July 1886 the pool cable companies held their semiannual meeting. The chairman of one of the companies proposed that the low rates were a boon and should be maintained for a time, since the eight cables were utilizing only 38 percent of their capacity. Increased traffic would be a service to the public while creating dividends for shareholders. But an Anglo-American Company shareholder, claiming to represent a million pounds sterling, insisted that if the Mackay-Bennett Company met the pool's low rate, they must reduce their rates more. "Company promoters should be taught," he said, "that there is no room for fresh cable schemers across the Atlantic." Critic Moore predicted that a row among the pool companies themselves was inevitable.

A few days later a report surfaced that quarreling had begun. At the annual general meeting of the French Atlantic Cable Company, a majority of stockholders present demanded the resignation of all board members. The chairman rebuffed the attempt to bring the proposal to a vote, and the membership refused to vote on any resolutions introduced by the board. The meeting had to be adjourned. The Tribunal of Commerce ordered the board to reconvene within two weeks, and the directors only maintained power through back-room maneuverings. Although plagued by squabbles, the pool companies were able to stay the course. Their officials believed that American public and business interests, knowing the value of a dollar, would come around. Moore suggested that Americans were shrewd enough to see through the ploy and that many businesses were paying more to use the Mackay-Bennett cable because they did not trust the pool companies.[11]

As the cable war continued, the construction of land telegraph lines across the United States was slowly progressing—too slowly for Mackay. In August 1886 George Ward visited Montreal to meet with the general manager of the Canadian Pacific Telegraph to confirm the connection between it and the Postal Telegraph Company. On August 10 lines were connected, joining Montreal and Richford, Vermont. Mackay followed up in September by going to Montreal to confer with officials of the Canadian Pacific Railroad. It was agreed that Postal Telegraph's connection from the East to the West Coast would be run along the Canadian Pacific railway, the largest telegraph system in Canada. Buying into the railroad company, Mackay became a director. A few months later Mackay purchased the Bay and Coast Telegraph, which ran from San Francisco to Santa Cruz. The Pacific Postal Telegraph Company was organized, and Henry Rosener was placed in charge. The Mackay company began to build lines along the Atchison, Topeka & Santa Fe Railroad lines into California and one line through Oregon and Washington.

Although Mackay's system eventually stretched from Vancouver to San Diego, it would be another fifteen years before the lines reached all major West Coast cities. As for the Atlantic Ocean cable enterprise, Mackay was convinced that he was pursuing the correct course. On September 20 the *Times* reported: "[Mackay] expresses himself very confident as to the result of the cable war."[12]

Rates stalemated, the cable contest continued into 1887. The pool companies, charging 12¢ a word, were losing money; the Mackay-Bennett Company, charging 25¢, was still supported by a segment of the business community. On April 20, 1887, general manager Ward announced: "As far as the Commercial is concerned, with its present support it can and will maintain the stand it has taken indefinitely." Unannounced was the fact that in some months the companies' deficits were made up out of Mackay's pocket.[13]

By early August rumors regarding reconciliation between the Gould faction and Mackay reached a fevered pitch. Mackay was returning to New York from Europe, and it was said that all parties were anxious to advance rates to a profitable basis. On August 8 Western Union stock rose and fell as reports that Mackay and Gould were in conference arose and were dispelled.

An interesting rumor surfaced in the *Times* "Wall Street Talk" column on the 9th: "One smart yarn that went circulating represented that Millionaire Mackay was in distress over the personal loss of a good many millions of dollars." The item continued that in his anger he threatened to smash the stock market to recoup his losses. While making no effort to influence the market, Mackay spent that day in his office. Writer John Russell Young met him and found him calm and resolute, but Mackay left for San Francisco abruptly on the six o'clock Chicago express.[14]

The reason for his abbreviated stay was that part of the smart yarn was true. While he was in England, his London banker had alerted him that the Nevada Bank had overdrawn its London account by 100,000 pounds. Mackay cabled peremptory questions, and the responses caused him to book immediate passage on a steamer. Upon his arrival he found Nevada Bank notes "kiting"—that is, drawn against insufficient funds—on the street. Wheat was irretrievably falling on the stock market, and men purporting to be his agents were buying at 30 percent over market value. He was facing immense personal losses. Moreover, the financial panic that seemed likely to ensue would mean disaster across the entire Pacific Coast. The crisis came to a head that Saturday, when the Nevada Bank closed for business at the regular midday hour. Cash on hand totaled $868. The bank was on the brink of collapse.[15]

Seeing the opportunity to bring down Mackay, Gould struck again. On August 19 the Gould-affiliated *New York Tribune* printed an exposé on its front page. The heading read "A California Wheat Deal." Subheads summarized the theme: "Flood's and Mackay's Vast Losses"; "From Six to Eight Millions Sunk"; "Wild Speculation That Narrowly Escaped Ending in a Great Panic." The lengthy piece offered a history of the Firm and the relationships among Flood, Mackay, and Fair. It noted that Fair had always claimed he made the fortune for the Firm and he had continued making millions since breaking away from his partners. He had left two years earlier, the report said, "because he could not get along with Mackay." The business capacity of Flood and Mackay, it stated, was greatly overrated. Without the inside information available to them in mining deals, they were consistently defeated. It continued that Flood's finances were threatened. He was suffering from Bright's disease and was unable to leave his country villa. Mackay, it said, was in worse financial straits than Flood. He had never been worth over $18 million, and half that sum was absorbed in his "unremunerative [*sic*]" cable and telegraph speculation. He had also lost heavily in a Texas railroad, previous wheat deals, and recent mine investments. Conjecture held that he felt the need of an audacious stock move to recoup previous losses.

The article said that the complex scheme, contrived by Flood and Mackay, was to begin by cornering the Chicago wheat market. Figureheads were used, so Flood and Mackay wouldn't be identified. When wheat reached $1 a bushel, it was reported, they would unload it, and with the $2,500,000 profit, boom all Comstock mines. By their creating an excitement in mining stocks, San Franciscans would buy and Flood and Mackay would sell, the two men clearing $5 million in that transaction. Finally, they would corner the whole supply of wheat in San Francisco and force it to new highs, making the "short" brokers agree to their terms. This last deal would bring in $6 million.

The article claimed the three deals would allow the "big manipulators" to "clean up" $13 to $14 million. Unfortunately for those involved, the price of wheat rose to the unprecedented price of $2.17 per cental (one hundred pounds), and after the buyers' resources were stretched as far as possible, the market broke. Prices plunged to $1.35, and the buyers lost millions. Asked in the article to say what he thought of Mackay's plight, ex-partner Fair did not hide his satisfaction. " 'Wa'al,' he remarked with that drawl which Mark Twain might envy, 'poor John, most of his money is now at the bottom of the Atlantic!' " Buried in the *Tribune* article was the statement: "Flood and his associates have declared over and over again that neither they in person nor the bank had any interest in the wheat transactions."[16]

The serious weakness in the premise of the *Tribune* story was the plausibility that either Flood or Mackay directed the scheme. Flood's health had been deteriorating for months and, as reported, he was largely confined to his home. Mackay had been in Europe since April, when the original wheat purchases were initiated, and his attention was occupied by his other businesses—cable, telegraph, and mining.

James G. Bennett's *Herald* labeled the article "A Stock-Jobbing Attack." Ostensibly authored by the *Tribune*'s correspondent, the *Herald* asserted it was the work of Gould, attempting to discredit Mackay so as to weaken Commercial Cable and Postal Telegraph. Years later Grant Smith remarked: "The scheme was to wreck the bank and Mackay and thereby end the cable and telegraph war. For mendacity and downright falsehoods that article cannot be surpassed." The *Herald* commented that it was disgusting and wearisome to again confront the endless round of Gould's trickeries and schemes. "Nobody would expect from Jay Gould a direct, open, and aboveboard attack; that is not his style. The tortuous and insidious attack on Mr. Mackay's credit is quite in Gould's old-time favorite mode of warfare. When hard pressed and driven into a corner he has always resorted to these tactics before yielding." But the attack had an effect. An unnamed Mackay associate was quoted in the *San Francisco Chronicle:* "It looks to me as if Mackay would be in the hands of Jay Gould before long, if, indeed, he is not there already."[17]

The fact was that many millions in Nevada Bank funds had been lost in a wheat scheme. The perpetrator was trusted bank vice president George Brander. With Mackay out of the country and Flood's health keeping him away, Brander had almost complete control of the institution. The bank, in previous years, had loaned heavily on wheat. It was understood that wheat loans were done because the bank owned a great warehouse at Port Costa and generally was advancing funds on grain already stored there.[18]

In April 1887 Brander contacted William Dresbach, a stock operator heavily interested in the wheat market, and several investors. All the wheat trades were conducted through John Rosenfeld, brother-in-law of the Rosener brothers, Mackay's old friends from Virginia City (Henry Rosener was managing the Pacific Postal Telegraph Company and was one of the limited investors in Commercial Cable). Although Rosenfeld gave no hint on whose behalf he was acting, there were suspicions that the Nevada Bank was the hidden power. The belief was that these were insignificant wheat brokers, and without the bank's backing, they could not have gained the needed credit. On April 26, 1887, a *New York Times* article had commented: "It is generally thought that the Nevada Bank combination has got the bears of the San Francisco wheat market in a tight place."

When the clique lost $400,000 thereafter, Brander began to rely on heavy loans to recoup the losses. The men thought to use the sheer weight of the bank's money to corner the market. Brander bought as long as Nevada Bank money held out, then he borrowed $300,000 from the Bank of California. In the end he also hypothecated private bonds. Mackay kept $1,000,000 in bonds for his wife in his private safe-deposit box. Brander stole them, along with those belonging to Flood's sister, amounting to a total of $600,000. He also used customers' collateral and securities. The wheat price advanced to $2.17 per cental, and only inexperienced and reckless operators could fail to see the impending collapse. On August 3, 1887, the bottom fell out of the market. When Flood called Brander to account and Brander confessed, Flood responded: "I ought to kill you."[19]

Mackay, harboring similar feelings, went directly to the bank the day he arrived from the East. He was intent on thrashing "the damned crook." A clerk described Mackay's arrival: "He strode through the office to the next room and burst out with, 'Brander, you god damned Scotch son-of-a ——.'" Brander averted a physical confrontation by remaining seated and making no reply. Finally, Mackay stomped back out. Mackay spoke to the press only to emphatically declare that neither he nor the bank owned any wheat. The problem, he said, was that Brander had loaned too heavily on wheat at too high a figure. "I knew nothing of these loans until I reached New York," he said. "It will take months to settle what these losses will amount to."[20]

Flood's first thought, when it was discovered that the bank had no liquid assets, was to place the bank in receivership. He had contacted leading San Francisco attorney Hall McAllister to draw up the papers. McAllister, at Flood's direction, visited the officers of the rival Bank of California to discuss the documents' details. Thomas Brown, the cashier who had held that position when the Bank of California had failed in 1875, protested that

McAllister had to dissuade Flood from allowing the Nevada Bank to fail. Brown insisted that the panic would send other houses to ruin.[21]

As a stopgap measure, the Bank of California loaned the Nevada Bank $1,000,000 on Mackay and Flood's personal notes. That Monday, when the Nevada Bank opened, the tried and true method of derailing a run was utilized—covering the counters with full coin trays. Customers, comforted by the sight of the trays, aborted the potential run, and business was conducted as usual.

When Mackay had arrived in San Francisco, he said to Flood: "Well, old man, it looks as though we might have to go back to work again." Although still in a weakened condition, Flood did what he could. Money was needed, so he reduced the price of his new block at Fourth and Market Streets for ready sale. When a firm was unresponsive to a request that it continue carrying a $500,000 loan, James Fair came forward offering to take the debt. One paper casually remarked: "James G. Fair had an idle million and a quarter that he had set aside to buy a ranch with, and this was placed at the bank's disposal." Probably unknown to Fair, his ex-wife also helped, loaning Mackay several millions in securities.[22]

For several weeks the wheat brokers, Rosenfeld and Dresbach, played for time. But on August 27, the twelfth anniversary of the collapse of the Bank of California, Mackay and the Nevada Bank repudiated them. The brokers issued a statement at the San Francisco Stock Exchange that they could not honor their contracts to purchase wheat. "A prolonged whistle from almost every part of the room greeted the . . . announcement." The losses fell upon other brokers, particularly those from the big houses, and although none failed, it was noted that the deal was one of the worst in the history of trade. Indignant traders claimed fraud and called the action a huge swindle. The failure had a disastrous effect on the mining stock market as well, and prices fell all around. The press now assailed Mackay and Flood, demanding an accounting of their roles in the debacle. The *New York Times* quoted an article calling Rosenfeld and Dresbach "straw men" and commented that some of Mackay's statements seemed to indicate that "the Nevada Bank has more intimate connection with the deal than its managers cared to admit."[23]

Within two weeks not only were the papers reporting that Mackay and Flood had not known about the wheat deal, they were calling Mackay's actions heroic. The *San Francisco Chronicle,* which, with the Gould-allied *New York Tribune,* took the lead in printing accusations against them, now remarked that Mackay "felt keenly the almost criminal imprudence of which Brander was guilty" and mentioned the "stormy scene" between the two upon Mackay's arrival in the city. On September 14 the *New York Times,* while saying that while "Mr. Flood has not been so sick a man as he has been desirous of making it appear, and that the truth is that Mr. Flood as a banker is not a success," acknowledged that Mackay "came on from London, and set to work to unravel the tangled web of bad financiering." It reported further that it was Mackay who forced Dresbach and Rosenfeld to admit their insolvency after they had been permitted to vacillate for weeks. Although creditors lost about a half million dollars, none was broken by the affair. "These losses were so widely scattered throughout the state among bankers, brokers, and farmers," it said, "that no individual loss led to insolvency." A friend later reported that Mackay "made good every loss."[24]

Fair commented: "Flood was a man who trusted everything to Brander and believed in him as he did in his God. I thought the man always fooled him. I believe there was a good deal of business Flood knew nothing about. . . . The manager began speculating and the bank did not know of it—that was the secret of it all. This man Dresbach was using their money all the time. . . . The critical time did not last a month and those terrible losses came within that time."[25]

The infusion of money loaned to the Nevada Bank allowed it to sell the accumulated wheat to best advantage. But prices remained severely depressed, and after it was sold and the appropriated bonds replaced, it was determined that close to $11,000,000 had been lost. (Economic historian Ira B. Cross thought the number was far greater, stating: "The attempted corner on wheat cost Flood and Mackay each approximately $8,000,000.") Between Flood's ill health and Mackay's problems with his communications investments, neither was in a position to rebuild the bank. On September 13, 1887, Flood, who had hoped to fill the position for the duration of his life, resigned as bank president. Brander was removed from all association with the institution. And, in a shocking turn, James Fair was named bank president.[26]

Although Flood had been instrumental in founding the bank and was devoted to it, the *San Francisco Chronicle* pointed out: "There is little prospect that his health will improve so as to again permit the close application to business for which he was famous." Still, it noted that Fair's appointment "was in the nature of a thunderbolt out of a clear sky." The San Francisco papers and much of the business community believed that Fair had invested a large enough sum to give him control of the institution. They saw it as his triumph over his old associates. Saying that the change was the sole topic of conversation in the business community, the *New York Times* commented on the masterly way Fair had "gained the upper hand of his former partners." The *Chronicle* remarked: "[Fair] has waited long for a chance to get even." But it went on to say that he was taking control as "an ally to his old partners." Mackay, who as majority partner had final say regarding Fair's taking control, believed it was the renewal of their partnership from the Comstock. They had had their differences, but Fair's coming to their assistance at a time when Mackay and Flood were financially embarrassed to Mackay meant simply, "the old firm was back in business."[27]

In taking control of the bank, Fair regained a public forum. "Others may go in for speculations if they please," he stated, "but a bank has no business to do so. The Nevada Bank got a little out of line. I suppose they had not round pieces in abundance. It is no secret that this bank was overloaning [*sic*] and that funds were running out faster than they could afford." Now he pledged that business would be conducted on more conservative principles and that only strict banking business would be done. Fair delivered on his promise. Under his presidency the bank regained public confidence almost immediately. In 1927 economic historian Cross pronounced it "one of the grandest feats of financiering on record." Fair served only two years, resigning the presidency on Oct. 9, 1889, to be succeeded by Flood's son, James L. Flood. The following year Mackay, along with the Flood and Fair interests, sold the greater part of their stock holdings to Isaias W. Hellman and associates, and the bank merged with the Wells Fargo Bank. Mackay remained a director of the combined firms until his death.[28]

As for those responsible for the disastrous venture, the discredited Dresbach assigned all his property for the benefit of creditors and was not heard from again. Brander, previously so renowned for conservative management traits that no one could explain his having approved the bank's heavy loans, sold his San Francisco property to Fair—presumably at tremendous loss— and departed for Europe. When things cooled off he returned, took over an ailing insurance company, was indicted on making false claims of the concern's assets, and again fled the country. Rosenfeld had lost all he had, signing $150,000 worth of property over to the bank (although his indebtedness was said to total well over $4 million). Mackay thought him an honest man who had been misled by Dresbach, and their friendship continued.[29] Devastated by the debacle, Flood sailed for Europe.[30]

Mackay was far from devastated. Even in the midst of dealing with the wheat issue, he kept his eye on Gould and the Atlantic cables. On August 18 he had authorized Ward to dispatch an announcement: "The war in cable rates has been going since May last year. In that time our customers have been paying 25 cents a word, while the pool companies have been charging only 12 cents a word. It is simple justice to our friends who stood by us steadfastly that we would adjust our rate to that of the pool companies." Beginning on September 15, 1887, the Commercial Cable Company reduced its rates to 12¢ a word.[31]

The same summer of 1887, Baltimore and Ohio owner Robert Garrett was suffering under tremendous mental strain. The ventures initiated by him were unsuccessful, costing $20,000,000—double what had been anticipated. The floating debt of the railroad was enormous, and a sinking-fund arrangement cost $750,000 a year. Garrett went to Europe to sell securities on the railroad in an attempt to raise $10,000,000. While he was gone the members of the syndicate who managed Baltimore and Ohio interests began retrenching by "summarily lopping off" Garrett's collateral enterprises. Garrett returned to the United States to find that, much against his wishes, J. P. Morgan had managed the purchase of the Baltimore and Ohio Telegraph Company by Jay Gould and Western Union. A fifty-year contract had been signed, paying the Baltimore and Ohio $5 million.

Under a front page headline "Swallowed Up At Last," the *New York Times* used a metaphor to report the usurpation of the Baltimore and Ohio: "The Western Union, that telegraphic anaconda, opened its jaws yesterday and took in the mouthful that has been waiting to be swallowed for many weeks." That evening at Stokes's Hoffman House, Garrett met with Stokes and a small group of friends, insisting for everyone in the restaurant to hear, "It's no trade gentlemen." He was indignant that the sale had been pursued while he, the president, was absent and unable to voice his opinion. He vowed to fight the sale. He had no intention of permitting Gould to capture his company "but intended to make another effort to combine all of Western Union's opponents in order to bring that monopoly to its knees." Although his statements caused a temporary depression of a couple of points in stocks of the two companies, it had no lasting effect. Baltimore and Ohio stocks had been plummeting for some time, and the company's directors had signed the contract of sale. Jay Gould merely smiled without comment at Garrett's complaints. The *Times* reported: "[Gould] was mightily tickled to know that his old rival was not politic enough to grind his teeth in the privacy of his own apartments."[32]

When Garrett suddenly left the city for Baltimore on October 9, Dr. Metcalf, his physician, accompanied him, and reports surfaced of mental illness. There were denials by friends, but on October 12 he resigned his presidency and never returned to public life. Shortly thereafter at the Maryland Club, he accused a group of men of being friends of Gould's, "glad that I have been sold out by him." A few days later, leaving the Camden train station to travel for his health, he shouted from his private car's rear platform, "Don't let Jay Gould steal this state of Maryland before I get back." Business failures and extreme aggravation at having been bested by Gould had caused a breakdown. He died at forty-nine years old in 1896, living the last several years of his life under physicians' care, having been pronounced insane.[33]

Stokes professed relief over the Baltimore and Ohio Telegraph Company's sale to the monopoly. He met the press with an air of calm, saying that Garrett's erratic rate cutting had affected United Lines and that it would be easier to fight one opponent than two. Since Mackay and Stokes's United Lines in large measure paralleled those of the usurped company, having forty thousand miles of line compared to the Baltimore and Ohio's forty-five thousand, Stokes claimed there would be no consequences for the Commercial Cable Company (which used United lines to transact its overland business).[34] But the truth was, to the extent that the deal strengthened Western Union, it weakened United Lines and Postal Telegraph.

At this time Albert Brown Chandler was serving as president of Postal Telegraph, directing its administrative affairs. He was an experienced career manager who, along with Mackay's longtime associate Henry Rosener, had developed Postal's coast-to-coast system. His résumé included working in the "President's Room," where Lincoln followed events during the Civil War, and a stint as president of the Atlantic and Pacific Line, as well as working at times for Western Union.

Now, wanting to abolish the practice of giving businesses rebates for using their lines, Chandler struck a deal. Confirming rumors published in the *New York Times,* on October 28, 1887, uniform rates were established for

Postal Telegraph and Western Union. Stokes's United Lines followed suit. Although franking privileges for government officials continued, rebates to businesses ended. The cutthroat competition in land telegraph was over.[35]

The agreement stabilized the market for all three companies. Mackay's lines, which ran from the Atlantic to the Pacific, would remain limited to certain areas. It was not until December 1888 that Postal pushed into New York City, and the following spring, with the incorporation of the South Atlantic Telegraphic Company, the line was linked to Baltimore. In the markets Postal served, it could compete because it boasted reliability and improved instruments—with four wires doing the work of sixteen ordinary ones. But, even with two hundred thousand miles of line by 1902, Postal never became a formidable rival to Western Union. It was handling only 17 percent of the United States' telegraph business in 1928 when it was sold. The inducements for Postal Telegraph in agreeing were improved profits with the elimination of rebates and a provision that allowed each company to use the other's equipment. This last guaranteed Mackay's international service would not be interrupted no matter where inside the states dispatches were destined.[36]

Gould and Western Union were the big winners in securing the agreement. The giant company would no longer consider buying out its opposition. Although Western Union's rates had been reduced along lines also serviced by its competitors and average tolls had decreased an average of seven cents, its dominant share of the market was now assured. And with rates now at a more reasonable level and a semblance of competition from Postal and United lines, charges of a monopoly were more easily refuted. Although from early 1890 through 1891 there was an effort in Congress to revive the idea of a federal telegraph service, government regulation would not occur until 1934.[37]

In part because of the cable and telegraph wars, in part because of other business pressures, in October 1887 Jay Gould was near collapse. On October 27 he and his family (his wife was also ill) sailed for Marseilles. He was gone five months. Cable prices remained at twelve cents a word. While he was gone the war, far from abating, escalated. The *Herald* called his return the most interesting in the history of piracy since Captain Kidd. The paper had been attacking him in absentia, urging indictments against him for a purportedly illegal railroad transaction. It now charged that his son had speculated against Gould's interests. Gould responded in an interview in the *New York Tribune*, saying: "The unprofitableness of the Mackay-Bennett cables is the chief cause for Mr. Bennett's animus against me." The *Herald* redoubled its efforts, publishing a cartoon with Gould's face on the body of a skunk. Five days later Gould distributed a vicious editorial attack on Bennett to all New York papers except the *Herald*. Without consulting Bennett, the editor procured a copy and ran the column the same day as the other papers, commenting: "The proprietor of the *Herald* lost his reputation long before Mr. Gould was ever heard of." Bennett was said to be quite pleased at his editor's performance.[38]

The rate war had severely depressed Gould's telegraph and cable earnings. Western Union's losses were estimated to be $750,000 a year. Gould revived attempts to get Mackay to raise cable rates to 60¢ and inquired about buying Mackay's interests. Mackay would not budge. Suddenly on July 11, 1888, it was announced that the cable war was ended. A Western Union director said: "Mr. Mackay and Mr. Gould have reached a full and satisfactory understanding with one another, and the fight that has cost lots of money all around is to be abandoned." An officer of Commercial Cable said: "Mr. Mackay was quite ready to join in an amicable arrangement for advancing the cable rates somewhat, not however, to anything like the old-time extortionate tariff." No matter the details, the news had an immediate, buoyant effect on Wall Street.[39]

It was not until July 30 that there was a formal agreement. In the meantime Mackay had to deny that he was selling his property to Gould. Even at news of the settlement, Ward had to be emphatic in denying that the Mackay-Bennett cable was to become part of the pool, saying all lines would engage in fair and open competition. Mackay's official statement was uncompromising: "I have tried twenty-five cents. I am content with that rate. The public is satisfied, and I will stay where I am." The *Tribune* reported: "The result is a unanimous agreement on the Commercial Cable rate of 25 cents a word." John Russell Young commented: "Peace was made by the absolute surrender of the monopoly."[40]

In the lengthy cable war, Mackay was the acknowledged winner. In 1894 Commercial Cable would lay a third Atlantic line (because the cables were owner financed, he could ignore the Panic of 1893 and the severe stock market depression that followed). In subsequent years he built a fourth line and a fifth—a heavy-duty cable renowned for its speed and the amount of traffic it could sustain. The other companies' pool agreement ended in 1895. At the cessation of the rate war in 1887, Gould was abashed. Month after month his companies had lost money. Now weary, he had to capitulate. Cable profits would continue to be significantly reduced. Although Gould's Western Union was now assured dominance in the telegraph battle, rates remained substantially lower, spurring public use, and over the next twelve years Western Union traffic increased 44 percent. Concerning the Atlantic cables, Gould was reported to say that there was no beating Mackay: "If he needs another million or two, he will go to his silver mines and dig it out."[41]

As on the Comstock, Mackay had taken the measure of a monopoly, this time withstanding Gould's cutthroat competitive tactics. Mackay's actions had averted federalization of the telegraph system while benefiting the public as well as the national and international business communities.[42]

CHAPTER 14

Losing Control

ON THE MORNING OF FEBRUARY 21, 1889, James Clair Flood died at the
Grand Hotel in Heidelberg, Germany. His personal history—coming from a
poor Irish family, working as a carpenter, a placer miner, and a saloon proprietor before earning one of the largest fortunes in the world—was an "only in
America" tale of the era. When he had sailed to Europe Fair described him
as "dying by inches." Almost blind and suffering from failing kidneys, he
went to Germany hoping specialists or the spas might help. He intended to
return to California but fell into a coma from which, over the last few weeks,
he could be roused only periodically.[1]

Flood's immense fortune was left to his widow, said to be a good-hearted
woman; his son, James L., who had already succeeded him in his enterprises; and his daughter. They carried on his tradition of liberal contributions to charities, notably at Christmas. Despite news editorials rebuking his
banking skills at the time of the wheat debacle, afterward pundits attributed

his failure to the overwhelming health issues that kept him from oversight duties. In reviews of Comstock finance by contemporaries, Flood was acknowledged as an astute stock-market operator and as having an incisive business mind. When his achievements are placed in historical perspective, the primary evidence of his ability is that he was the most successful and longest in power of all Comstock mining operatives.[2]

Flood's long illness did not lessen the impact of his death on Mackay, who grieved over the loss of his friend. Speaking of him, Mackay said: "In all that goes toward the development of manhood, the best man I have ever known."[3]

About the time of Flood's death, Mark Twain asked J. P. Jones and Mackay to help finance the Paige typesetting machine that he was promoting. At first they were interested, but later they became disinclined to invest. Perhaps due to Mackay's being caught up with the sale of the Nevada Bank of San Francisco or perhaps because of diminishing returns in the mines—by 1893 the entire workforce in Comstock mining was composed of 259 people—in July 1890 Joe Goodman wrote to Twain that both Mackay and Jones had become "somewhat diffident in the matter of huge capitalization." Jones, in mid-February 1891, backed out altogether, explaining that he did so because Mackay and others had decided not to invest.[4]

A further reason Mackay lost interest may have been the purchase in December 1890 of a London mansion for Louise. It had been built by the third Duke of Leinster and had been recently renovated by a millionaire who, because of financial reversals, was forced to sell. The mansion had a ballroom opening on a broad terrace that overlooked St. James Park. The *San Francisco Examiner* commented: "[Mrs. Mackay] saw the Carlton Terrace house, liked it and Mr. Mackay saw that she wanted it, upon which, of course, Mr. Mackay bought it."[5]

A few years earlier, on March 23, 1886, Louise had had a dream come true when she and her mother were presented to Queen Victoria at Buckingham Palace. Also presented that afternoon was one of the old-time Comstockers, C. W. Bonynge, and his daughter. Bonynge's wife was to have been presented as well, but shortly before the event a London newspaper revealed that she had been divorced and so was ineligible. Bonynge believed Louise or John, who was in America at the time, to be involved in the release of the report.[6]

Bonynge had been an English soldier in the Crimean War and served as one of the six hundred commemorated in Tennyson's "The Charge of The

Light Brigade." Coming to America, he had been a stockbroker in Virginia City in the 1860s before moving to San Francisco. The Bonynge-Mackay grudge went back to that time. In 1876 he had been one of the brokers who joined Keene in the "bear" attack on the Con. Virginia. Just before moving to Europe in 1878, he placed articles in the *San Francisco Chronicle* attacking the Firm as using various schemes to milk the public.[7]

After Louise's presentation to the queen she held a dinner and reception, omitting the Bonynges, while hosting, among many others, the Prince and Princess of Wales. The *New York Times* commented: "The triumph of Mrs. Mackay is complete. She will soon forget all about her Meissonier misfortune and the Arc de Triomphe. . . . His Royal Highness has solemnly consecrated by his presence the new temple of American hospitality in Buckingham Gate, and the fête of last week was a signal success. . . . The crowd of ambassadors, ambassadresses, peers, peeresses, and American citizens soon became bewildering."[8]

A few days after the event the *Manchester Examiner and Times* and the *Echo* in London printed articles of a decidedly different nature: "It is not generally known that Mrs. Mackay, who entertained the Prince of Wales on Wednesday night, and whose parties will be conspicuous this season, was once what the Americans call a washwoman, what we call a washerwoman. She was a poor woman, with two children to support, and washed clothes for some of Mackay's miners out in Nevada." A similar article at about the same time described how—although she now hosted royalty, appeared at the opera, and wore the most fashionable of clothes—eighteen years ago Mrs. Mackay had kept a boarding house in Virginia City. A crude circular issued another fallacious version of her life on the Comstock and included the boorish declaration that "Mrs. M . . . with all those other seekers after bonanzas sought relief in the sage bushes instead of a jasper or alabaster lined, and rose of ottar scented closet." Someone collected these items and sent them anonymously to Louise Mackay and everyone else who had attended her dinner.[9]

Louise filed a lawsuit against the *Examiner and Times,* and on December 5, 1889, the decision was rendered in her favor. She felt compelled to bring the suit for libel because the account they printed was false, but upon winning, she requested only that the paper apologize, pay costs, and contribute a large sum to a charity that she named. The judge commented that the defendants were lucky Mrs. Mackay was willing to be so lenient.[10]

Mackay was not as forgiving as his wife. He hired detectives to identify who had been responsible and offered two hundred pounds for prosecutable evidence. None was forthcoming. Still, Mackay believed he knew. Two years later he said: "Long ago I suspected Bonynge was the instigator of certain vile attacks upon Mrs. Mackay, which appeared in papers in New York and London. Having proof that this was true, I determined to punish him the first time I met him."[11]

They met in September 1891 in the Nevada Bank offices. The *New York Tribune* called the ensuing row "a regular rough and tumble one, in which Mackay had decidedly the best of it." Bonynge received a black eye and contusions of the face, while Mackay was said to have only a few scratches. Mackay got in the first two punches, apparently catching Bonynge by surprise, and Bonynge never gained the offensive. Half a dozen clerks rushed in to separate the men. Mackay remarked: "The sound thrashing he received he well deserved for circulating these stories about Mrs. Mackay. I'm not so handy with my fists as I used to be twenty-five years ago up on the Comstock, but I have a little fight left in me yet and will allow no man to malign me or mine."

The *New York Tribune* reported: "There is small chance of the quarrel being carried any further as Bonynge is afraid of Mackay." The San Francisco papers carried ironic accounts of the battle of the elderly combatants: Mackay was nearly sixty years old, Bonynge a couple of years older. Ambrose Bierce memorialized it with a mock-heroic poem, said not to be one of his best efforts.[12]

Mackay's relationship with Edward Stokes also went awry in the early 1890s. Stokes had acted for Mackay in purchasing the telegraph receiver's certificates in 1885. On January 13, 1892, he sued Mackay for $75,000, claiming that he was not merely an agent, as Mackay maintained, but rather a partner. Stokes argued that Mackay was to be repaid out of their telegraph profits and the surplus was to be divided between them. When disputes arose between the two, an agent for Mackay negotiated the transfer of all the telegraph property for $100,000. A contract was signed and $25,000 was paid. Stokes now asked for the balance plus interest. Mackay said he had never authorized the agreement; his agent said he was tricked into signing it. Between 1884 and 1888 Mackay had advanced Stokes $1,233,338. When an accounting was demanded, Stokes insisted that the agent first sign the contract. The agent signed, later maintaining it was a mere matter of form. After the original trial and finding for Stokes, there were appeals, reversals, and new findings. Finally the contract was found to be valid, and Mackay was forced to pay the balance and interest. Stokes lived another six years, almost all of it mired in litigious disputes involving other adversaries.[13]

Another of Mackay's publicized associations was with the heavyweight boxing champion Gentleman Jim Corbett. As a young man Corbett, working at his father's livery stable in San Francisco, got a job as a messenger for the Nevada Bank. He rose to the position of assistant receiving teller. During that time he sparred frequently after work at both the stable and the bank. A small gymnasium had been built in the basement, and when Corbett worked out there Mackay took note of his skill. Corbett began training at the Olympic Club, where he went on to win the middleweight and then the heavyweight club titles. When a match was arranged with professional Jack Burke, and upon hearing that Burke was getting paid one hundred dollars, the bank gave Corbett a week's vacation to train. There was no decision in the Burke fight, although Corbett held his own for all of its eight rounds. Corbett began winning other fights, left the bank, and soon turned to professional pugilism.[14]

Mackay boxed himself for exercise, but he was disdainful of it as a quasi-profession. The rules of the day specified bare knuckles, and the fight went on until one combatant could no longer continue. Corbett, on his way to meet a famous fighter named Kilrain, described running into Mackay at a train station:

"Is this Jimmie Corbett?" he asked. . . ."What in the world are you doing here?"

"I have left the Bank and am a pugilist now," I informed him a little proudly.

"A pugilist—huh!" he retorted. "And where are you going now?"

"Down to New Orleans to fight Jake Kilrain."

"Well," he said, "I hope you get a damn good licking!"

Corbett defeated Kilrain, and Mackay was won over. In 1892 Gentleman Jim knocked out John L. Sullivan for the world heavyweight championship, the first fought under the Marquess of Queensberry Rules, featuring five-ounce gloves and three-minute rounds—although no specified number of rounds. Corbett knocked Sullivan out in the twenty-first. Mackay, in London, gathered many notables with him and had his New York office cable him the round-by-round account. Two years later, when Corbett was in New York preparing to travel to England to play the lead in a melodrama called "Gentleman Jack," Mackay sent for him. "During my call Mr. Mackay said that in England 'they look on a pugilist as a pretty tough customer; and I want to give you some letters to some of my friends over there to show them that a man can be both a pugilist and a gentleman.'" Mackay dictated the letters, they visited pleasantly for a couple of hours, and Corbett went on his way. Once in London, Corbett was visited by Mackay's son Clarence. His father had cabled him from America, instructing him to show Corbett a good time. Corbett commented that Clarence did as his father requested.[15]

Mackay's loyalty manifested itself in many ways. Whenever Mackay stayed in San Francisco, Alexander O'Grady, who lived with the Mackay family in Paris, was invited for a visit. O'Grady reported: "Invariably there would be an envelope on the table, to which [Mackay] would direct my attention. It was not his way to say what was in it, but I always found some large bills, sometimes as much as $500. I was a struggling young lawyer in those days, and that money meant much to my mother and me."

Edmond Godchaux, a longtime friend who had traveled with Mackay to Europe three times, asked Mackay for a $100,000 loan for his uncles' business. Mackay wrote the check, but when Godchaux examined his uncles' books he tore it up. The uncles cajoled and pleaded until Godchaux agreed to ask again. They offered goods in a warehouse to Mackay as collateral. Mackay wrote a new check, telling Godchaux to keep the goods for himself and his sisters, as they might need it—and they did.[16]

John Russell Young said Mackay lived "an honorable and blameless life." C. C. Goodwin said that when summoned by the judgment angel, Mackay could say, "I gained many millions, but I kept my hands clean." Rollin Daggett wrote a poem dedicated to Mackay when both men were in their sixties. The work fondly recalls the old days on the Comstock and toward its conclusion reads:

> Though prizes were but few, John, (the wheel
>   was not to blame,)
> And largely fell to you, John, we all enjoyed
>   the game.
> But gold was only part, John, of Fortune's gifts
>   devout;

You drew a sunny heart, John, to keep wealth's
    mildew out.[17]

There are countless other testimonials from associates and beneficiaries of Mackay's magnanimity. The far-reaching consensus as to his character and honesty did not prevent Mackay's being seriously wounded in an assassination attempt.

Behind the Lick House, the first of San Francisco's great hotels, there was a paved court that ran from Sutter to Post Street, known locally as Lick Alley. There were large iron gates at each end, but they were never closed. Mackay reported that at noon on February 24, 1893:

> I was walking down Sutter-st., [sic] bound for my room to pack my valise, as I intended to go to Virginia City this afternoon. When I reached Lick Alley I thought it would be a short cut to the hotel. I had hardly taken two steps into Lick Alley when I heard a pistol shot. I paused for a moment, rather startled, for I had not seen any one in the alley when I entered it, and did not at first know from which direction the report of the pistol came. Suddenly a voice exclaimed: "Mr. Mackay, you have been shot," and looking up I saw the gentleman who brought me to my room.
>
> "No, I am not shot," I replied.
>
> "Yes you are," he insisted, "for I saw the dust fly from your coat when the bullet struck you."
>
> I put my hand around under my coat and, sure enough, when I looked, my hand was covered with blood. I then felt pain in my back. It all happened in a second, and, turning half way around I saw my assailant standing at the head of the alley with his pistol pointed at me. I then hurried across to the east side of the alley, and as I did so I saw the old man put a pistol, as I thought, to his mouth, and fire it again. He then reeled and fell, and I continued up the alley until we reached Mr. Bonner's cart, and he drove me to the Palace Hotel.[18]

John Bonner, who managed a florist shop in Lick Alley, had just stepped out of his wagon when he heard the loud report.

"[Mackay] was perfectly cool, and he walked toward where I was standing, oblivious to the fact that his would-be murderer was standing close behind him with a pistol leveled at him.

"It all happened in a flash. The rough looking man did not fire at his intended victim again, but, turning the weapon about, pointed it at his own breast and fired. He whirled about in the street and fell to the pavement."

Bonner drove Mackay to a doctor's office, but the doctor was not in. He then took him to the Palace Hotel, to his room on the first floor, and a surgeon was called. The *New York Times* described the intense excitement: "It was noon, and thousands of people were on the streets in the business quarter where the shooting occurred. They crowded to the scene of the tragedy, and the narrow alleyway was soon black with curious sightseers." At the hotel it was again reported that Mackay was unruffled: moving about his apartments telling friends that he did not know the man who shot him.

The surgeon removed the bullet, which had entered below the shoulder blade, struck a vertebra, and lodged near the spine. As Mackay rested, the doctor announced that the wound might not be fatal, but that the patient was by no means out of danger.[19]

The would-be assassin was seventy-year-old W. C. Rippey. He survived his suicide attempt, was hospitalized, and was later tried and convicted. The day of the shooting Mackay said he believed Rippey must be a crank, and the first reports in the papers described him as such. One account said he was perpetrating an "insane attempt." Another correctly identified him as a ruined speculator.[20]

Sometime earlier the Firm had promoted a Utah mine, and Rippey had invested in it. When his investment was lost, he spoke with Flood at the Bank, who told him the mine had not paid expenses for years. This caused him to think they had deliberately allowed him to lose his money. He brooded and concluded Mackay was "the head and backbone of the crowd" that caused his troubles. "They drove me to it," he said. But these statements were made as he recuperated a month after the incident. At the time another motive seemed to prompt his action: a note addressed to the *San Francisco Examiner* was found in his pocket. "Food for reflection. Paid $150,000 for sapphire to place on the jaded person of his wife, a sufficient amount to have saved 500 of his paupers from a suicidal grave. Just think of it. Inscribe it on his tomb."[21]

There was a wave of populist sentiment in the United States in the 1880s and early 1890s. The Knights of Labor and other workers' unions, Marx's International, Farmers' Alliances, immigrants, socialists, anarchists, prominent artists and writers were among the disaffected that protested during the era. Violent strikes, riots, and the hanging of four anarchists for the bombing death of seven policemen in the Hay-market Riot in Chicago (although none of the convicted was identified as having thrown the bomb) were the most obvious manifestations of the period. Teddy Roosevelt identified the time as "a riot of individualistic materialism, under which complete freedom for the individual . . . turned out in practice to mean perfect freedom for the strong to wrong the weak." In the election of 1892, the populist candidate for president received over a million votes and carried four states. Rippey's note to the press reflects the undercurrent of resentment to the excesses of those who prospered in the age and the desperation of those trammeled by the era's unfettered individualism. He later told the press: "Mackay never injured me individually, but he was the main figure of the crowd that ruined so many men."[22]

During the next few weeks, Mackay was confined to his rooms. He was cheerful to the guests who were admitted, but his condition was guarded. R. V. Dey was kept busy replying to cablegrams and telegrams. Typical of the messages was one from the Comstock: "Your friends of the Comstock, which include a large majority of the population, congratulate you on your providential escape from the bullet of the assassin and hope for your speedy recovery." In mid-March his condition took a turn for the worse. He suffered an attack of appendicitis—a serious ailment in the era. Inflammation in the abdomen restricted his diet, and he became dangerously weak. Absolute silence was kept. Only nurses keeping vigil, the doctors, and attendants were allowed to see him. His condition was thought to be critical, and it was feared that a difficult surgery would be required. By March 21 the crisis had passed, and it was reported that he improved hourly and was out of danger.[23]

After the assassination attempt, there had been confusion in cabling Louise regarding John's condition. She had just returned to London from Italy, where her daughter, Eva, was having marital problems. A cable from John told her of the shooting. It was a slight wound, he said, and there was absolutely no cause for alarm. For the next two weeks, cables arrived almost daily telling her of his improvement. Then there was no word for several days. When she cabled inquiring if there was a problem, Dey wrote saying John was recovering and there was no cause for alarm. Something was not right. Why had John not responded himself? She cabled an urgent message to longtime friend Sam Rosener, who responded immediately that John was dangerously ill from "an inflammation of the appendix vermiformis." As soon as could be arranged, she and her son Clarence sailed for America.

During the voyage, word arrived from John that the inflammation had sub-sided and the danger was past. The doctor sent a cablegram as well, saying that her husband was much improved.[24]

In New York John Jr., who was now serving as a director on the board of Commercial Cable and had been in France looking into cable affairs, met his mother and brother. On April 4 the three left by train to cross the coun-try. Once in San Francisco, seeing that John was improved, the family vis-ited friends and vacationed as John recuperated. Louise believed that John had tried to write her at the onset of the appendix attack but Dey had not sent the message. Instead, Dey had sent his cable that there was no cause for alarm. She believed Dey hated her, having lost stature because she had turned John's attention from the West to New York and Europe. She con-jectured that because his entire life was lived vicariously through John, Dey wished to be the principal mourner if her husband died. She did not reveal her suspicions, and Dey remained his secretary the remainder of Mackay's life.[25]

In mid-May the family traveled east to Chicago and New York. Although John conducted some business, and those dealing with him said he seemed as active as always, most of his time was spent with Louise and the boys. At the end of the month he accompanied Louise and Clarence to the vessel on which they returned to Europe, and he and John Jr. returned to business. Two months later Mackay again suffered an attack of appendicitis. After his doctor diagnosed the problem, a second physician was consulted. They treated Mackay for several days then decided an operation was necessary. It was successfully performed at his rooms at the Belgravia Apartment House. On August 11 the *Times* reported that Mackay was improving. The doc-tors' bills had been exorbitant, and Mackay balked at paying. But by the time Alice O'Grady, Mrs. Mackay's ex-house manager, whom Mackay never ceased to regard as a member of the family, visited he could joke about it. She asked if his appendix had been removed. "I don't know Alice," he said, "but from the size of the Doctor's [*sic*] bill he must have removed all of my organs."[26]

Although Mackay was recovering, news on another front was disturbing. Mackay's stepdaughter, Eva, the Princess of Colonna, had separated from her husband. Eva, who at one time was rumored to be considering taking the veil of a nun, had followed her half-sister, Ada, in becoming among the first Americans to secure a foreign title through marriage. A substantial number of the wealthiest American daughters would follow. By the end of the century it was estimated that two hundred million dollars was exported to Europe through such matches. Eva's husband, Don Ferdinand Colonna, was handsome but rakish, and he gambled. Three years before the marriage, at age twenty-one, he had lost a small fortune given him by his grandmother, and his debts had to be settled by an uncle. Before marrying Eva, he had assured the Mackays that his gambling was a thing of the past. It had not been. For a time the marriage seemed successful; there were three children. But in eight years the princess had been forced to pawn her jewels several times and accept large contributions from Mackay to pay her husband's debts. These payments were in addition to the annual two-hundred-thousand-dollar income supplied by Mackay.[27]

When Colonna won at the table he wasted large sums, often for necklaces and other jewelry that the princess never saw. Despite such insulting behavior she stayed with him, hoping he would change. There were instances of violence, including Colonna's bruising his wife's cheek and throwing a wine bottle at her. Colonna and Louise Mackay hated one another. In the final scene between them, she told him what she thought of his actions, and he compared his princely birth with her origins. When husband and wife separated in the fall of 1893, Prince Colonna hired a detective to follow the princess. She reported that Colonna proposed to the detective that he abduct their oldest son to force his wife to return to him. Learning of the scheme, she left for America.[28]

In New York in February 1894, she received word that the French courts had denied her divorce and ordered the children be placed in custody of Colonna's aunt and uncle. Shortly thereafter, accompanied by John Mackay Jr., she took the children to California. A year and a half later, cut off from Mackay's wealth, Colonna finally consented to a legal separation giving his wife custody of the children. The settlement included a stipend of twelve thousand dollars a year to be paid the prince. Princess Colonna raised her children and then devoted herself to charity, becoming the head of the Red Cross in Rome. She died in Santa Margharita, Italy, in March 1919 from influenza contracted while doing Red Cross work.[29]

With his stepdaughter in California, John Jr. traveling, and the rest of the family in Europe, Mackay was again living by himself in New York. He broke his solitude by entertaining guests. In December 1893 Mark Twain wrote to his wife about dining at Mackay's with a few other "old gray Pacificcoasters." They talked of their "gypsying" days on the Comstock: "Indeed it was a talk of the dead. Mainly that. . . . For there were no cares in that life, no aches and pains, and not time enough in the day (and three-fourths of the night) to work off one's surplus vigor and energy." Twain told his wife that Mackay had a pet monkey—a "winning little devil" that caused nothing but trouble for the servants.[30] In another letter a few months later, Twain spoke of his very public business failures and an assignment he had made to benefit his creditors. He wrote that Mackay called on him, saying, "Don't let it disturb you, Sam—we all have to do it, at one time or another; it's nothing to be ashamed of."[31]

Although John and Louise's relationship was congenial, the light of his life was his first son, John Jr., known to intimates as Willie. All of Mackay's plans and business ambitions included him. Bright, with a winning personality, John Jr. was a graduate of Oxford University. He resembled his father in many ways but had an easy, outgoing personality that Mackay Sr. lacked. Mindful of his own preoccupation with business, Mackay Sr. encouraged a pleasure-loving, active lifestyle for the young man. Still, Mackay Jr. did not aspire to a life of ease. He believed that men of wealth should take an active part in public affairs. One of his ambitions was to represent Nevada in Congress. He had many friends in America as well as among English nobility. A companion described him as "one of the kindest-hearted, truest, open-handed, charitable of men and most loyal of friends." He continued, "Willie's equal would not be found among the many rich young men of the country."[32]

Mackay Sr. wanted only the best for his son. When John Jr. became romantically involved with an older woman who was a teacher, his father was displeased. Mackay Sr. remarked to a friend, "Why, she is only a school teacher." The friend bristled. "So was my wife," he said. Mackay reconsidered then responded, "Well, so was mine." Young Mackay gave a dinner at the Waldorf in 1893 in honor of Consuelo Vanderbilt. It was said to be one of the most sumptuous affairs ever given at the restaurant, and rumors of a pending marriage spread. They proved false. The two were merely good friends, and he was exhibiting his gregarious and generous nature.[33]

Since 1890 John Jr. had been involved in his father's businesses in Europe. He had a quick mind and an aptitude for commerce and had served as a director and member of the executive committees of the cable and telegraph companies. He was always ready to assist other young men he thought worthy in obtaining an education or a position suitable to them. On May 24, 1894, he gave a speech at a banquet for officials of the Postal Telegraph Company and the Commercial Cable Company. The talk, in part, outlined his business and leadership philosophy: "Whatever I may undertake in telegraph affairs will be right and clean from top to bottom, with a due appreciation of the work of every man, whether by my side or in the struggle of rivalry. . . . I have a fellow-interest with you all. If my word or influence is needed to assist you in your work, rest assured it is yours."[34]

In October 1895 young Mackay and two friends were at a country chateau he had rented at Mange, a remote rural district in France. Mackay Sr. had permitted him to extend his vacation an extra month. John Jr. and his friends had recently bought horses, and they practiced jumping hurdles and ditches on a track he laid out. Young Mackay was an excellent rider and had driven teams from an early age. Once his father saw him handling a "four-in-hand" and compared him to a renowned Comstock teamster, exclaiming: "That boy can drive as well as Curly Bill!" John Jr. had talked extensively about the time when he would own a racing stable. On October 18 the young men held a race. The night before a friend had a nightmare, seeing a rider wearing colors like young Mackay thrown and killed. He told his friends of the dream at lunch, but the others gave it no credence.[35]

The horse John Jr. chose to ride was restive. His friend had ridden it the day before and recommended he ride another. Young Mackay persisted. As they ran, the horse bolted from the track into thickets. John Jr. managed to avoid several obstacles but, losing control, reeled in the saddle and was thrown headlong into a tree. The wounds were ghastly; both his eyes were crushed. He was carried to the house on a mattress. One report said he remained unconscious for three hours before doctors revived him, another that while being carried he told his friends that if they were tired they should put him down and rest. When he was placed in bed, his dog lay at his feet refusing to move.

John Mackay Sr. had gone to dinner in San Francisco with two business associates. He went to his office afterward and there received a dispatch from son Clarence in Paris saying that John Jr. was dangerously ill. Two more telegrams were received, apparently preparing the father for the worst. In the early morning a fourth message arrived: "Willie was thrown from a horse yesterday, and never recovered consciousness. He died last evening."[36]

CHAPTER 15

Universal Eulogy

AT JOHN MACKAY JR.'S funeral, there were two wreaths of mauve orchids and white lilacs. One was inscribed: "From a broken-hearted mother"; the other: "From a broken-hearted father." The lives of Louise, in Europe, and John, in San Francisco, had been ruptured. John sat in his room moaning "Oh, God, what have I done to deserve this?" The few friends allowed to see him were afraid he was losing his mind. One commented, "He was crushed to earth by [John Jr.'s] death." When he left for New York, a friend who accompanied him said Mackay scarcely said a word. The friend said it was the saddest trip of his life. When John and Louise came together for the services, John would barely mention the death. Anything else he said came slowly and painfully. They planned a $300,000 mausoleum at Greenwood Cemetery on Long Island and arranged that Masses be said in perpetuity. On February 12, 1896, Louise returned to Europe. She lived in seclusion many months, not resuming social activities for two years.[1]

When he was able, John's daily routine continued as before, but he never fully recovered. Chandos Fulton, a friend, commented: "The smile leaving his manly features [was] succeeded by a serious, almost mournful, expression, which became habitual." When those who knew John Jr. visited Mackay's Fifth Avenue apartment, he would show them his son's room, unchanged since his death. He would point out various articles that held meaning until emotion overwhelmed him. Another friend, Edward C. Platt, said: "Mr. Mackay was never the same man after the death of his son John. . . . Sometimes he might seem to forget it for two or three months. Then something would arise to recall it and his changed demeanor plainly showed that the deep wound was still open." When feeling disheartened, Mackay would ask someone like James E. Walsh, secretary of the Flood estate, to walk with him. Often the walks ended at the cemetery where Mackay's old-time friends lay.[2]

Through the grief Mackay continued attending to business. Commercial Cable absorbed Postal Telegraph. Over two million dollars was expended constructing the twenty-one-story Commercial Cable Building in New York City, and a new Atlantic Ocean cable was laid. In this last venture Mackay contributed all funding, as James Bennett, although maintaining his ¼ interest, did not participate. Clarence Mackay, untrained for business, replaced the older brother. It was a lonely and difficult task. Mackay Sr. rarely sought companionship, insulating himself with silence. On May 17, 1898, Clarence married Katherine Duer, only child of the New York society Duers. Ironically, having long before lost her desire for it, Louise Mackay was now connected to the New York society set. She reacted, after attending the stateside wedding, by returning to Europe. Mrs. Clarence Mackay began to host dinners and entertainments the equal of those her mother-in-law had given in Paris and London.[3]

About that time doctors found lesions in Mackay's heart. They warned him against physical activity or business that would put a strain on him. It depressed him further to have to slow down and not be able to take stairs two at a time.[4]

When he retired from the Comstock mines in 1894 his projected fortune had been $25,000,000, making Mackay one of the richest men in America. But he kept his great wealth active, and even through his years of despair, he served on various companies' boards of directors. Besides his cable businesses and directorship of the Canadian Pacific Railroad, he served as vice president of the Spreckels Sugar Company, director of the Southern Pacific Railroad Company, and director of a new railway to be constructed from Havana to Santiago, Cuba. He was the largest investor in a copper company in Idaho, near a town that his mining superintendent named Mackay. He maintained large interests in a mining company at Nome, Alaska, and the Sprague Elevator and Electrical Works of New York. He owned large amounts of property in California, including the Grand Opera House and half the Nevada Block in San Francisco. After his death estimates in the newspapers projected his wealth at from $50 million to $100 million. Personal secretary Dey said that although his businesses were in order, "I don't suppose Mr. Mackay himself knew within $20,000,000 of what he was worth."[5]

Going against management's usual mantra that competition requires strict economy regarding labor costs, Mackay insisted wages remain high. In months when the telegraph provided no profits, Mackay made up the difference. When a supervisor suggested they could save $12,000 a year by cutting salaries 10 percent, Mackay proposed cutting the supervisor's wage, and if at the end of the year the supervisor found it workable, they would institute the reduction as policy. At the end of 1899, after $3,300,000 worth of additional Commercial Cable stock was issued at $150 a share, Mackay encouraged employees to subscribe on the same terms as the stockholders—one of the first American companies to do so.[6]

Ten years earlier Andrew Carnegie had issued his "Gospel" on wealth. He argued that progress resulted from competition and that the few who accumulate the most capital should live as examples in homes that exemplify the highest refinements of civilization. He continued that as capitalists compete against each other in manufacturing, the quality of commodities improves, costs decline, and "the poor enjoy what the rich could not before afford." To this point his philosophy mirrored that of most of the era's wealthy entrepreneurs. But his philosophy included a radical proposal. He suggested that when large estates were left at death, they be heavily taxed. He argued that this would induce the affluent to administer their surplus wealth during their lifetime to benefit the community: "to place within its reach the ladders upon which the aspiring can rise." Carnegie, explaining his own public endowment, which has carried forward into the twenty-first century, said that rather than leave great sums to heirs, which was often harmful to the recipients, the few should provide funding for parks, works of art, and institutions to improve the general condition of the people.[7]

Mackay's philosophy on wealth, as evinced by lavish spending on Louise on the one hand and his lifetime of altruism on the other, seems in concert with the crux of Carnegie's doctrine. Toward the end of his life he asked his friend Godchaux, the city recorder of San Francisco, for a list of the most worthy charitable institutions in the city so that he might create a philanthropic organization to help distribute his wealth, but Mackay died before implementing a plan.[8]

Still, Mackay maintained his person-to-person munificence. A friend told of walking with Mackay on Sixth Avenue when a shabbily dressed old Comstocker slouched toward them trying to avoid notice. "I've been looking for you for some time," Mackay greeted the man. "Looking over an account book of mine I found I had borrowed some $300 from you going on to forty years ago. . . . Here's $50 on account, and come and see me tomorrow, and I'll pay you the balance." As Mackay and his friend moved on, Mackay commented that the man was too proud to beg or borrow. The man had done him many favors on the Comstock, he said, and "I am glad to do one for him now."[9]

Another anecdote told of a former Comstock merchant named Fleishacker who had fallen on hard times and was summoned to the Nevada Bank. The merchant's son, Mortimer, accompanied his father. Mackay said he had bought some stock for the father and had sold it at a profit of $20,000. He produced that amount in gold, and the men carried it away. Telling friends that he had bought stock in their names so he might give them money was a device Mackay employed repeatedly.[10]

P. H. Lannen, an associate, said: "His charities I know amounted to more than a quarter of a million dollars a year." In 1902 Mackay told a confidante that he was going to sell a San Francisco lot at Third and Market Streets for $1,250,000 so he might make a gift of $75,000 to each of two old-time friends under the guise of a commission. It was a matter of comment that when he was in San Francisco Mackay carried thousands of dollars in large bills. He never said why, but it was assumed that it was to help needy old friends.[11]

There was one time when Mackay did not respond kindly when he was asked for aid. Godchaux told of a woman who walked into Mackay's private office at the Nevada Bank requesting money. Godchaux was shocked when Mackay rebuked her sternly and ordered her from his room—but Mackay then stepped into the bank and asked a clerk to follow her and find out if she was truly needy.[12]

Dey told of watching Mackay tear up over $1,200,000 in notes owed him. "That fellow's broke," he'd say, and throw notes from $500 to $50,000 into the wastebasket. "That fellow's dead, Dey . . . can't collect from him, so here goes." And he would tear up the next paper. Ripping up another, he said, "Now this fellow may get on his feet one of these days, and if he does I guess he'll remember the amount; but in the meantime I might as well get the evidence out of the way so that he won't be pressed in case anything happens to me."[13]

William Guard, later publicity director of the Metropolitan Opera Company, was sent by the *New York Herald* to interview Mackay and ask if he would be helping Maurice Grau organize the new opera company. Mackay responded: "Though I seldom talk to newspapers, I authorize [you] to tell your editor that John W. Mackay is a friend of Maurice Grau, believes in him, and that Maurice Grau can have anything he wants." Grau began the new company with $150,000 capital. How much Mackay contributed no one knew, but it was understood that it was a considerable percentage.[14]

When James Walsh and his wife visited New York in 1900, Mackay was courteous and secured opera tickets for them but would not go with them. He told them he had not attended any place of amusement since John Jr.'s death five years earlier. At that time the Met, under Grau's exceptional management, hosted the greatest performers in the world. Toward the end of the Walsh's visit, they finally persuaded Mackay to attend a performance with them. The opera may well have been conducted by the renowned Luigi Mancinelli, who led over five hundred recitals at the Met in his career, in the year 1900 including performances of *Faust, Aida, Carmen, Roměo et Juliette,* and *Don Giovanni.* Mackay told the Walshes that he enjoyed the performance, saying he felt better for having gone.[15]

After the turn of the century, he regained some of his equilibrium. In 1901 he planned the next phase of his cable system. As early as 1870, cable builders had considered the idea of a Pacific Ocean cable, but the magnitude of the project, complicated by the great depth of the water, frightened off potential investors. For a dozen years presidents sent messages to Congress advocating legislation that would provide for telegraphic communications across the Pacific. The navy had done depth soundings preparing for a government project. Sometime after August 1898, when the Philippines came under the control of the United States, Clarence had suggested building such a cable to his father. Mackay had been intrigued and now, at age seventy, disregarding his doctors' orders about strenuous activity, he began to work on a plan.[16]

On August 22, 1901, Mackay sent a cablegram to John Hay, the secretary of State: "Sir I beg leave to state I wish to lay and operate a submarine cable or cables from California to the Philippine Islands, by way of the Hawaiian Islands, by means of an American corporation to be organized hereafter. I ask no guarantee in connection with the same." The answer would not be given until the submission of a formal application. The bureaucratic requirements included a written proposal of terms and conditions on which the permission was requested, a copy of the corporation charter, "and such other information as will aid the Department in reaching a decision." Additionally, since the navy would want reasonable rates for transmitting messages in lieu of privileges granted to this company, Mackay was directed to place a detailed scheme before the Department of the Navy.[17]

Once the formal application and lengthy statements were completed, the *New York Herald* supplied the public with details. The cable would be completed in two years. The government would have no proprietary interest, although in time of war the cable would be turned over to the appropriate government departments.

Owing to widespread support for Mackay and respect for his business acumen, there was little question of the project's approval. The secretary of the American Asiatic Association, representing 250 businesses, wrote that they desperately wanted a cable and that allowing Commercial Cable Company to build it was the quickest and best way to achieve that end. A California member of the House of Representatives noted: "Mr. John W. Mackay, the head of this company, is held in the highest esteem in California, both as a businessman and a citizen." Senator George C. Perkins of California, who had previously introduced a bill authorizing the government to build a Pacific cable, asked that his bill be set aside. He noted that Mackay would take special pride as an American in building the first Pacific cable and that the Commercial Cable Company would use its own money and assume all risks. The members of the Senate Committee on Naval Affairs were impressed with the proposal. They endorsed it and voted unanimously to indefinitely table the Perkins bill.[18]

Although Mackay spoke personally to President Theodore Roosevelt about the project, gaining his full support, there were complications with the naval ocean soundings' being put at Mackay's disposal. As always, Mackay ignored the problem and pushed ahead. When work commenced, it was without government subsidy or guarantee of any kind.[19]

Mackay was at sea, on his way to Honolulu, when he heard of the death of Alexander O'Grady's mother, Alice, who had tended the Mackay children in Paris. Mackay telegraphed Dey, telling him to make all the funeral arrangements at Mackay's expense and to preserve the body until he might return.[20]

In May 1902, back in San Francisco, Mackay announced that the American terminus for the Pacific cable would be built in the city. He hoped to complete the section of the cable to Hawaii by Thanksgiving. At age seventy-one, despite bouts of gout and rheumatism, his health was good as were his spirits. But a friend commented, "That old springing step was gone."[21]

By mid-July 1902 Mackay had returned to London. His business career was coming to an end. Three ships would leave England shortly, carrying the first sections of the Pacific cable. On Tuesday, the 15th, Mackay was in the London office. He sent a lengthy cable to the company lawyer in New York regarding the delay in receiving the Pacific soundings and then went to lunch with Ward. As they left, Mackay told the longtime manager: "I'll lay that Pacific cable, and then retire from business."

During lunch Mackay said he felt ill. He had chills, although it was the hottest day of the summer. Ward took him home. Doctors were summoned. His heart condition had worsened, and one lung was filled with fluid. Over the next few days he seemed to improve, then he relapsed. On Saturday Louise contacted a priest, who issued last rites. In the early evening on Sunday, July 20, John Mackay fell unconscious and died.

Mackay's quote "There is no law in mining but the point of the pick" suggests much about the way he lived his life. He worked hard, following his bent without allowing outside entities to deter him. He had little interest in politics, was involved in no reform movement, and—although fighting against oppressive monopolies—did not right the wrongs of the system; his strategies involved forming competing giant companies. His legacy, sustained into the twenty-first century, is that he was generous and helped those around him, worked in the public interest, and succeeded when standing up to the most powerful titans of the age.

At his death tributes were paid on both sides of the Atlantic and across America. The lengthy obituaries, tributes, and resolutions of regret were strikingly similar, reflecting a universal eulogy. Although discussing his personal wealth, his achievements in mining, and his facilitation of commercial and public telegraph and cable use, they emphasized his philanthropy and goodness. They featured interviews with associates who, without exception, honored the man. Commentary in the *New York Herald* began with the quote "When a good man dies he begins to live." It called him a gentleman "in the truest sense of the word" and went on to: "[Mackay's] countless benefactions, the many people he cheered in dark hours, the unfortunate ones he set on their feet . . . will be heard from in the future, and his good deeds told and retold as the years roll by." The *San Francisco Examiner* proclaimed: "He was manly, he was kind and he was honorable." The *New York Times* remarked, "Although born on foreign soil there were few men who so genuinely illustrated in their appearance and their character the spirit and form of the Western Nation's ideal."[22]

Clarence would carry on the business, which at its merger with International Telephone & Telegraph in 1928 consisted of 73,004 miles of transatlantic and transpacific ocean cable, 386,093 miles of land telegraph lines, and boasted, "The sun never sets on the Mackay System." Louise withdrew from public life. She and her son gave permanent gifts to the University of Nevada in John Mackay's name.

A statue of him stands on the Reno campus quad, below Mackay Stadium, in front of the Mackay School of Mines. On June 10, 1908, a state holiday was declared, and thousands viewed the statue's dedication.[23] The artist was Gutzon Borglum, whose many sculptures and monuments include the faces on Mount Rushmore. The Mackay piece is a heroic portrayal: dressed in miner's clothes, he gazes into the distance. One hand clutches a piece of ore, perhaps representing all he accomplished. The other hand is atop the handle of a pick, and his shirtsleeves are rolled up, as if he were ready to begin again. The larger-than-life depiction illustrates the comment of R. V. Dey, his secretary and friend of forty-two years: "There never was another like him, even in the big days when men seemed to have a chance to be bigger than they are nowadays. He was a stalwart."[24]

Notes

PROLOGUE

1. Sam Davis, quoted in *San Francisco Examiner,* July 21, 1902.

2. Wells Drury, *An Editor on the Comstock Lode* (1936; reprint, Reno: University of Nevada Press, 1984), 62–63.

3. A primary area of New History study is the environment, allowing us to begin to understand the mutual exchange that occurs between humans and their physical settings. It directly contradicts the "clean money" philosophy of the earlier age, emphasizing the tremendous toll mining took on wildlife, forests, and water resources. Patricia Limerick commented: "The miners left; their works remain" (Patricia Nelson Limerick, *The Legacy of Conquest: The Unbroken Past of the American West* [New York: W. W. Norton & Co., 1987], 18). See also William Cronon, *Changes in the Land: Indians, Colonists, and the Ecology of New England* (New York: Hill and Wang, 1983); William G. Robbins, *Colony and Empire: The Capitalist Transformation of the American West* (Lawrence: University Press of Kansas, 1994).

4. Sam P. Davis, *The History of Nevada* (Reno: Elms Publishing, 1913), vol. 1, 414.

5. Ida M. Tarbell, *The Nationalizing of Business, 1878–1898,* A History of American Life, vol. 9 (1927; reprint, New York: Macmillan Co., 1969), 40; John Russell Young, *Men and Memories,* ed. May D. Russell Young (New York: F Tennyson Neely, 1901), 449–50.

6. John Russell Young, *Men and Memories,* 445–46. For a discussion of interven-

tions being proposed in the U.S. Senate just prior to Mackay's involvement, see *New York Herald,* January 18, 1884.

7. Smith's material on Mackay is at The Bancroft Library, University of California, Berkeley, referenced as: Smith, Grant, "Manuscripts and Notes Chiefly Concerning the Comstock Lode," BANC MSS P-G 244, cartons 2 and 2a.

8. *San Francisco Examiner,* July 21, 1902.

9. Amelia Ransome Neville, *The Fantastic City: Memoirs of the Social and Romantic Life of Old San Francisco,* ed. and rev. Virginia Brastow (Boston and New York: Houghton Mifflin Co., 1932), 189–90.

10. Oscar Lewis, *Silver Kings: The Lives and Times of Mackay, Fair, Flood, and O'Brien, Lords of the Nevada Comstock Lode* (New York: Alfred A Knopf, 1947), 67; Robert Gracey and Baruch Pride, quoted in Smith, "Manuscripts and notes," carton 2a, "John W. Mackay, Bonanza King," 6–7; William Miller, *A New History of the United States* (New York: George Braziller, 1958), 246.

11. Drury, *An Editor,* 217–19; *New York Herald,* July 27, 1902.

12. *New York Herald,* July 27, 1902.

13. See Alfred D. Chandler Jr., *The Visible Hand: The Managerial Revolution in American Business* (Cambridge: Harvard University Press, 1977); William G. Roy, *Socializing Capital: The Rise of the Large Industrial Corporation in America* (Princeton, N.J.: Princeton University Press, 1997). To describe what he means by power, Roy begins with Max Weber's definition: "The ability of one actor to impose his or her will on another despite resistance." He broadens it to include "the extent to which the behavior of one person is explained in terms of the behavior of another." And Roy adds a second dimension, "structural power," defining it as: "The ability to determine the context within which decisions are made by affecting the consequences of one alternative over another" (13).

14. Smith, "Manuscripts and notes," carton 2a, "John W. Mackay, Bonanza King," 312–13.

15. Ibid., 303, 307; Neville, *Fantastic City,* 189.

16. *San Francisco Examiner,* July 21, 1902; Smith, "Manuscripts and notes," carton 2, notebook titled "Comstock Notes," section Q, 2.

17. William E. Sharon, R. V. Dey, in Smith, "Manuscripts and notes," carton 2a, "John W. Mackay, Bonanza King," 322. The description of the two comes from C. C. Goodwin, *As I Remember Them* (Salt Lake City: Salt Lake Commercial Club, 1913), 126, 161–62. Goodwin's version of the incident is somewhat different, with Mackay losing his temper as well.

CHAPTER 1. *Starting Out*

1. Smith, "Manuscripts and notes," carton 2a, "John W. Mackay, Bonanza King," 3, 312–13; *San Francisco Chronicle,* July 21, 1902; *New York Tribune,* February 22, 1885; Ellin Berlin, *Silver Platter* (Garden City, New York: Doubleday & Co., 1957), 195–96.

2. Smith, "Manuscripts and notes," carton 2a, "John W. Mackay, Bonanza King."

3. Research is inconclusive with regard to Mackay's father's religion.

3. *New York Times*, July 21, 1902.

4. Smith, "Manuscripts and notes," carton 2a, "John W. Mackay, Bonanza King," 3–4.

5. Smith, "Manuscripts and notes," carton 2, "Comstock Notes," 3.

6. J. C. Furnas, *The Americans: A Social History of the United States 1587–1914* (New York: G. P. Putnam's Sons, 1969), 288, 382.

7. *Bureau of the Census, Historical Statistics of the United States: Colonial Times to 1970, Part 1* (Washington D.C.: Government Printing Office, 1975) Series C, 106; Samuel Eliot Morison and Henry Steele Commager, *The Growth of the American Republic* (1930; rev. ed., New York: Oxford University Press, 1958), vol. 1, 500–502. English demonization of the Irish dates to Richard II's seeking to conquer them in 1395, condemning them as "savage Irish, our enemies." In the sixteenth and seventeenth centuries on the English stage, Irish were portrayed as wild men. American colonists described them as slothful, barbarous, and wicked. See Ronald Takaki, *A Different Mirror: A History of Multicultural America* (Boston: Little, Brown & Co., 1993), 25–29.

8. *New York World*, July 27, 1902, in Smith, "Manuscripts and notes," carton 2a, "John W. Mackay, Bonanza King," 3.

9. John S. Hittell, "The Mining Excitements of California," *Overland Monthly and Out West Magazine* 2, no. 5 (May 1869): 413.

10. The obituary in the *San Francisco Chronicle*, July 21, 1902, states his voyage was by way of the horn; the *San Francisco Examiner* of the same date says he arrived via Panama. This latter is repeated in most biographical sources.

11. R. Guy McClellan, *The Golden State: A History of the Region West of the Rocky Mountains; Embracing California, Oregon, Nevada, Utah, Arizona, Idaho, Washington Territory, British Columbia, and Alaska, from the Earliest Period to the Present Time . . .* (Philadelphia: William Flint & Co., 1872), 137.

12. See Harry L. Wells, "Gold Lake: The First Stampede in the California Mines," *Overland Monthly and Out West Magazine* 4, no. 23 (November 1884): 519–25; and George and Bliss Hinkle, *Sierra-Nevada Lakes* (1949; reprint, Reno: University of Nevada Press, 1987), 89–101.

13. John S. Hittell, "Mining Excitements of California," 414.

14. Hinkle and Hinkle, *Sierra-Nevada Lakes*, 108.

15. Rollin M. Daggett, "The Bonanza King," *San Francisco Chronicle*, April 3, 1898.

16. Charles Howard Shinn, "California Mining Camps," *Overland Monthly and Out West Magazine* 4, no. 20 (August 1884): 175.

17. "Pioneer Days of California," *Overland Monthly and Out West Magazine* 8, no. 5 (May 1872): 458; Malcolm J. Rohrbough, "Mining and the Nineteenth-century West," in *A Companion to the American West*, ed. William Deverell (Malden, Mass.: Blackwell Publishing, 2004), 117, 123. For a discussion of race relations in the West,

see Sarah Deutsh, "Landscape of Enclaves," in *Under an Open Sky: Rethinking America's Western Past,* ed. William Cronon, George Miles, and Jay Gitlin (New York: W. W. Norton & Co., 1992).

18. William S. Greever, *The Bonanza West: The Story of the Western Mining Rushes 1848–1900* (Norman: University of Oklahoma Press, 1963), 49; Oscar Lewis, *High Sierra Country* (1955; reprint, Reno: University of Nevada Press, 1984), 109–11; Titus Fey Cronise, *The Natural Wealth of California* (San Francisco: H. H. Bancroft & Co., 1868), 230.

19. Hinkle and Hinkle, *Sierra-Nevada Lakes,* 217–18.

20. *London Daily Telegraph,* July 21, 1902.

21. William Downie, "Hunting for Gold," in *Gold Rush: A Literary Exploration,* ed. Michael Kowalewski (Berkeley: Heyday Books, 1997), 197.

22. Ibid., 218–19; "Pioneer Days of California," 460–61; Walton Bean, *California: An Interpretive History* (New York: McGraw-Hill Book Co., 1968), 141–42, 234; Berlin, *Silver Platter,* 51–52.

23. Hinkle and Hinkle, *Sierra-Nevada Lakes,* 220; Smith, "Manuscripts and notes," carton 2a, "John W. Mackay, Bonanza King," 7; Cronise, *Natural Wealth of California,* 232.

24. *San Francisco Call,* July 21, 1902; John S. Hittell, in Smith, "Manuscripts and notes," carton 2a, "John W. Mackay, Bonanza King," 4; Smith, "Manuscripts and notes," carton 2a, "John W. Mackay, Bonanza King," 6–7, 35. Much of the information came from Robert Gracey, who knew Mackay in Sierra County, and E. C. Bradley, vice president of Mackay's Postal Telegraph Company; Robert L. Fulton, "Reminiscences of Nevada," *First Biennial Report of the Nevada Historical Society, 1907–1908* 1 (1909): 85.

25. Berlin, *Silver Platter,* 21, 42, 61–63; *San Francisco Call,* July 21, 1902; Smith, "Manuscripts and notes," carton 2a, "John W. Mackay, Bonanza King," 302; *San Francisco Chronicle,* July 21, 1902.

26. Dan DeQuille, *The Big Bonanza* (1876; reprint, Las Vegas: Nevada Publications, 1974), 24–25, 31–33; Ronald M. James, *The Roar and the Silence: A History of Virginia City and the Comstock Lode* (Reno: University of Nevada Press, 1998), 8.

CHAPTER 2. *The Fountainhead*

1. DeQuille, *Big Bonanza,* 106–7; *Annual Mining Review and Stock* Ledger (San Francisco: Verdenal, Harrison, Murphy & Co., 1876), 13.

2. *San Francisco Call,* July 7, 1902.

3. DeQuille, *Big Bonanza,* 31–32.

4. *Sacramento Union,* July 8, 1861, from *Territorial Enterprise,* quoted in Grant H. Smith, *The History of the Comstock Lode: 1850–1920* (1943; reprint, Reno: University of Nevada, 1998), 14.

5. *Annual Mining Review and Stock Ledger,* 15. Dan DeQuille told the same story in 1876 as well, saying it came from H. T. P. Comstock (*Big Bonanza,* 32). Historian Ronald James points out that this, like other early-day stories, cannot be confirmed. Casting the earliest miners as "devil-may-care drunks" rather than serious miners fit easier into the image of the legendary Wild West. Collected by writers with literary aspirations, the tales survive into this century as local history, demonstrating the power of myth in reinforcing the "Wild West." Ronald James, *Roar and the Silence,* 18–20.

6. George Thomas Marye Jr., *From '49 to '83 in California and Nevada: Chapters from the Life of George Thomas Marye, a Pioneer of '49* (San Francisco: A. M. Robertson, 1923), 109.

7. *Annual Mining Review and Stock Ledger,* 14; Smith, "Manuscripts and notes," carton 2a, "John W. Mackay, Bonanza King," 36; *San Francisco Chronicle,* July 27, 1902.

8. *Annual Mining Review and Stock Ledger,* 14; *San Francisco Chronicle,* July 27, 1902; Eliot Lord, *Comstock Mining and Miners* (1883; reprint, Berkeley: Howell-North, 1959), 64; *Territorial Enterprise,* June 25, 1884.

9. Lord, *Comstock Mining and Miners,* 64; *Territorial Enterprise,* June 25, 1884.

10. Lord, *Comstock Mining and Miners,* 64–67.

11. J. Ross Browne, Carson City, U.T., Apr. 5, 1860, to Lucy [Browne], Washington D.C., in *J. Ross Browne: His Letters, Journals and Writings,* ed. Lina Fergusson Browne (Albuquerque: University of New Mexico Press, 1969), 235. Browne had been the official reporter of the California Constitutional Convention in 1849. He traveled extensively before and after his Nevada sojourns, gaining renown writing and illustrating for *Harper's Magazine* and producing books based on his adventures. The tone of much of his writing was humorous, and his perceptions were accurate.

12. Lord, *Comstock Mining and Miners,* 67–71; Ronald James, *Roar and the Silence,* 39–41; Rohrbough, "Mining and the Nineteenth-century West," in Deverell, *Companion,* 123. The most comprehensive treatment of the war is Ferol Egan, *Sand in a Whirlwind: The Paiute Indian War of 1860* (New York: Doubleday & Co., 1972).

13. Smith, *History of the Comstock Lode,* 21–22.

14. *San Francisco Call,* July 21, 1902; Ronald James, *Roar and the Silence,* 36–37.

15. *San Francisco Call,* July 21, 1902; DeQuille, *Big Bonanza,* 375.

16. Lord, *Comstock Mining and Miners,* 62, 79–80, 90; Rohrbough, "Mining and the Nineteenth-century West," in Deverell, *Companion,* 113.

17. Smith, *History of the Comstock Lode,* 24; Lord, *Comstock Mining and Miners,* 89, 216.

18. Myron Angel, ed., *History of Nevada* (1881; reprint, Berkeley: Howell-North Books, 1958), 573–74; DeQuille, *Big Bonanza,* 234; Smith, *History of the Comstock Lode,* 83; Ronald James, *Roar and the Silence,* 36–37.

19. Lord, *Comstock Mining and Miners,* 64–67, 72, 94; E. B. Scott, *The Saga of*

Lake Tahoe (Crystal Bay, Lake Tahoe, Calif.: Sierra-Tahoe Publishing Co., 1957), 366, 372.

20. J. Ross Browne, *A Peep at Washoe and Washoe Revisited* (1864, 1869; reprint, Balboa Island, Calif.: Paisano Press, 1959), 179–81.

21. Smith, *History of the Comstock Lode,* 23. Smith says that perhaps as many as 10,000 people came that summer. In *Roar and the Silence* Ronald James, while noting that thousands came and went, uses the federal census of 1860 that gave the population as 3,017 and a territorial project that documented 4,437 two years later (37).

22. Lord, *Comstock Mining and Miners,* 94–96; Rollin Daggett, in *San Francisco Examiner,* January 22, 1892.

23. Mrs. Frank Leslie, *California: A Pleasure Trip from Gotham to the Golden Gate, April, May, June* (1877; reprint, Nieuwkoop, Netherlands: B. De Graaf, 1972), 277.

24. Henry T. Williams, *The Pacific Tourist* (New York: Henry T. Williams, 1877), 210; Ronald James, *Roar and the Silence,* 94, 177–80; Leslie, *California: A Pleasure Trip,* 280; *Annual Mining Review and Stock Ledger,* 15; *San Francisco Examiner,* January 22, 1892; Drury, *An Editor,* 17.

25. Richard E. Lingenfelter, *The Hardrock Miners: A History of the Mining Labor Movement in the American West 1863–1893* (Berkeley: University of California Press, 1974), 3–4; Smith, "Manuscripts and notes," carton 2a, "John W. Mackay, Bonanza King," 35. Regarding the $25,000 goal quote, William E. Sharon, mine superinten-dent and nephew of Senator William Sharon, told Smith: "Sixteen years later. When [Mackay's] income from the Consolidated Virginia bonanza exceeded $300,000 a month, Jack [O'Brien] good-naturedly twitted him about that remark. To which Mackay replied with his slow smile; 'Well, I've ch-changed my m-mind.'"

26. Sam Davis, *History of Nevada,* vol. 2, 1063, 981; James W. Hulse, *The Nevada Adventure: A History* (1965; 6th ed., Reno: University of Nevada Press, 1990), 96.

27. Smith, *History of the Comstock Lode,* 80–83, 94–98.

28. Ibid., 84–90.

29. Lord, *Comstock Mining and Miners,* 415.

30. Smith, "Manuscripts and notes," carton 2a, "John W. Mackay, Bonanza King," 62–63.

31. Smith, *History of the Comstock Lode,* 45–46; Lord, *Comstock Mining and Miners,* 389; Lingenfelter, *Hardrock Miners,* 15–16, 23. The heat estimate comes from G. F. Becker's notable government report, quoted in Smith, "Manuscripts and notes," carton 2, "Comstock Notes," n.p.

32. Smith, "Manuscripts and notes," carton 2a, "John W. Mackay, Bonanza King," 292.

33. Drury, *An Editor,* 15–16; *San Francisco Examiner,* January 22, 1902.

34. Ronald James, *Roar and the Silence,* 167, 186; Smith, "Manuscripts and notes," carton 2, "Comstock Notes," section R, n.p.; *Virginia Evening Chronicle,* December 28, 1874; Drury, *An Editor,* 135–36.

35. Lord, *Comstock Mining and Miners,* 302.

36. Lewis Coe, *The Telegraph: A History of Morse's Invention and Its Predecessors in the United States* (Jefferson, N.C.: McGarland & Co., 1993), 90; *San Francisco Chronicle,* July 21, 1902; Smith, "Manuscripts and notes," carton 2, "Comstock Notes," section R, n.p.; Sam Davis, in *San Francisco Examiner,* July 21, 1902.

37. D. O. Mills, in *New York Herald,* July 21, 1902; Smith, *History of the Comstock Lode,* 103.

38. Smith, *History of the Comstock Lode,* 103–4; Sam Davis, *History of Nevada,* vol. 2, 1063; Marye, *From '49 to '83,* 110.

39. Smith, *History of the Comstock Lode,* 48–49.

40. Ibid., 105–6; Marye, *From '49 to '83,* 109.

41. Lewis, *Silver Kings,* 69; George Lyman, *The Saga of the Comstock Lode* (1934; New York: Ballantine Books, 1971), 262–63; Smith, *History of the Comstock Lode,* 105–106. Grant Smith attributes Lyman's story to Mackay's friend R. L. Fulton, who said it took place in 1863. Smith points out that the Kentuck deed, made to J. M. Walker, was dated June 1865 (see Smith, "Manuscripts and notes," carton 2, "Comstock Notes," n.p.).

42. Smith, *History of the Comstock Lode,* 91, 105–6; Lord, *Comstock Mining and Miners,* 225.

43. James E. Walsh to Grant Smith, September 12, 1929, in Smith, "Manuscripts and notes," carton 2, "Comstock Notes," section S, n.p.

44. Smith, *History of the Comstock Lode,* 91, 105; Cecil G. Tilton, *William Chapman Ralston, Courageous Builder* (Boston: Christopher Publishing House, 1935), 229; Sam Davis, *History of Nevada,* vol. 2, 1063; *New York Times,* February 22, 1889.

45. Smith, *History of the Comstock Lode,* 118; *San Francisco Examiner,* December 30, 1894.

46. *San Francisco Examiner,* December 30, 1894; Sam Davis, *History of Nevada,* vol. 1, 398; Lewis, *Silver Kings,* 116.

47. *San Francisco Chronicle,* December 30, 1894; Lewis, *Silver Kings,* 125; Henry R. Whitehill, *Biennial Report of the State Mineralogist of the State of Nevada for the Year 1875 and 1876* (Carson City: John J. Hill, state printer, 1876), 124. Correspondence reveals Fair modifying machinery and the workings of his mine influencing others. In 1872, as superintendent of the Hale & Norcross, he complained that wheels on the incline cars were wearing out within a month, so he wanted to try a new design. He wrote a foundry manager asking that eight wheels, designed by a third man, be cast in a manner Fair and the manager had discussed. "Do not bore the wheels out as I wish to fit them to the axles here," he directed. In another letter he told a superintendent in Amador County that he would send a diagram of the Hale & Norcross safety cage if the man would pay the fee for a draftsman or photographer. "Nevada Mining Companies collection, 1862–1901," letterbook 3, Hale & Norcross, March 1872–July 1873, 12, 41, The Huntington Library, San Marino, California.

48. *San Francisco Examiner*, December 30, 1894.

49. Ibid.; Smith, *History of the Comstock Lode*, 116; Lewis, *Silver Kings*, 126–31.

50. "Mark Twain's Letters From Washington," number II, December 16, 1867, http://www.twainquotes.com/18680107t.html (accessed April 6, 2005); "J.M. Walker, Republican Delegate to Republican Convention 1868," http://www.political graveyard.com/bio/walker5.html (accessed April 6, 2005); Tilton, *William Chapman Ralston*, 229; *Reno Gazette*, November 28, 1936, in Smith, "Manuscripts and notes," carton 2, "Comstock Notes," section T, n.p.

51. Smith, *History of the Comstock Lode*, 104–5; Lord, *Comstock Mining and Miners*, 281.

52. Smith, *History of the Comstock Lode*, 117, 119n.; Tilton, *William Chapman Ralston*, 229; Lewis, *Silver Kings*, 43.

53. Marye, *From '49 to '83*, 111; Smith, *History of the Comstock Lode*, 145, 150n.; *San Francisco Chronicle*, September 14, 1887; John Russell Young, *Men and Memories*, 443.

54. Oscar Lewis believed the friendship merely showed that Fair believed Mackay might prove useful to him someday. Lewis, *Silver Kings*, 129–30.

CHAPTER 3. *Hale & Norcross*

1. Berlin, *Silver Platter*, 21, 62–63, 91–96.

2. Ibid., 100–102.

3. Lewis, *Silver Kings*, 73; Sharon Lowe, "Opium in Comstock Society," in *Comstock Women: The Making of a Mining Community*, ed. Ronald M. James and C. Elizabeth Raymond (Reno: University of Nevada Press, 1998), 102–4.

4. Berlin, *Silver Platter*, 110–16, 128.

5. Lewis, *Silver Kings*, 98; Andria Daley Taylor, "Girls of the Golden West," *Comstock Women*, ed. James and Raymond, 279; Berlin, *Silver Platter*, 137; Mrs. Robert Howland, on Hirschman, in Smith, "Manuscripts and notes," carton 2, "Comstock Notes," notebook s, n.p.

6. Berlin, *Silver Platter*, 110, 133–34; *San Francisco Chronicle*, July 21, 1902; Mrs. Howland to Smith, May 24, 1930, in Smith, "Manuscripts and notes," carton 2, "Comstock Notes," notebook s, n.p.

7. Berlin, *Silver Platter*, 140, 146–54; Smith, "Manuscripts and notes," carton 2a, "John W. Mackay, Bonanza King," 305. It was Mrs. Mooney's statement that Mackay and Fair owned a mine down the canyon and Grant Smith's deduction that it was the Occidental.

8. Smith, "Manuscripts and notes," carton 2a, "John W. Mackay, Bonanza King," 298.

9. Book of Deeds, Storey County Records, Nevada, book 27, 710; Smith, "Manuscripts and notes," carton 2, notebook titled "Mackay. W Nev. Bank Postal(?)

Co, Cable Cos. Posted 6," section s, n.p.; Mrs. M. M. Mathews, *Ten Years in Nevada, or Life on the Pacific Coast* (1880; reprint, Lincoln: University of Nebraska Press, 1985), 39–43.

10. Smith, "Manuscripts and notes," carton 2a, "John W. Mackay, Bonanza King," 310–11.

11. Smith, "Manuscripts and notes," carton 2, notebook 1b, 310–11.

12. Lord, *Comstock Mining and Miners*, 246–48; Smith, *History of the Comstock Lode*, 49–51.

13. Smith, *History of the Comstock Lode*, 91; *Virginia Evening Chronicle*, October 20, 1874.

14. See Grace Dangberg, *Carson Valley: Historical Sketches of Nevada's First Settlement* (1972; reprint, Reno: Carson Valley Historical Society, 1979), 17–26; John M. Townley, "Reclamation in Nevada 1850–1904" (PhD diss., University of Nevada, Reno, 1976), 208–11.

15. Angel, *History of Nevada*, 124; Hubert Howe Bancroft, *The Works of Hubert H. Bancroft*, vol. 25, *History of Nevada, Colorado and Wyoming 1540–1888*, (San Francisco: History Co., 1890), 238.

16. *The Peoples Tribune*, January 1870.

17. F. E. Fisk, quoted in Angel, *History of Nevada*, 290–91; Sam Davis, *History of Nevada*, vol. 1, 414.

18. Smith, *History of the Comstock Lode*, 58, 116.

19. Lord, *Comstock Mining and Miners*, 288–89.

20. Smith, *History of the Comstock Lode*, 116–17; Lord, *Comstock Mining and Miners*, 289, 303. Lord listed the stock price at $1,300 per foot on January 8.

21. In *Comstock Mining and Miners*, Eliot Lord said: "Though apparently a desperate gambler, he took no ill-considered risks, for he never professed to see further into the Lode than could be seen by the aid of pick and drill" (302).

22. Tilton, *William Chapman Ralston*, 231; George Lyman, *Ralston's Ring*, (1937; reprint, New York: Ballantine Books, Comstock Ed., 1971), 131–32; *San Francisco Call*, November 14, 1885, reported that the widow was paid eight thousand dollars a foot.

23. *Gold Hill News*, February 27, 1869.

24. *San Francisco Examiner*, December 30, 1894.

25. Smith, "Manuscripts and notes," carton 2a, "John W. Mackay, Bonanza King," 324.

26. Smith, *History of the Comstock Lode*, 118.

27. Marye, *From '49 to '83*, 112–13.

28. Smith, *History of the Comstock Lode*, 264; Smith, "Manuscripts and notes," carton 2, notebook titled "Mackay duplicates and discards 8," section M, n.p.; George Littleton Upshur, quoted in Lewis, *Silver Kings*, 234.

29. Smith, *History of the Comstock Lode*, 264.

30. Lord, *Comstock Mining and Miners*, 303–4; DeQuille, *Big Bonanza*, 405;

Charles Howard Shinn, *The Story of the Mine: As Illustrated by the Great Comstock Lode of Nevada* (1910; reprint, Reno: University of Nevada Press, 1980), 176–77.

31. Richard E. Lingenfelter, with and introduction by David F. Myrick, *1858–1958 The Newspapers of Nevada: A History and Bibliography* (San Francisco: John Howell—Books, 1964), 52–53; Ronald James, *Roar and the Silence*, 230. In February 1872, when Lynch died, Doten borrowed ten thousand dollars from Sharon to gain ownership. Thereafter the *News* became Sharon's primary adherent and apologist in Nevada. Alf Doten, *The Journals of Alfred Doten: 1849–1903*, ed. Walter Van Tilburg Clark (Reno: University of Nevada Press, 1973) vol. 2, 1157.

32. *Gold Hill News*, September 29, 1869.

33. Lord, *Comstock Mining and Miners*, 305; Smith, *History of the Comstock Lode*, 119.

34. Miriam Michelson, *The Wonderlode of Silver and Gold* (Boston: Stratford, 1934), 230–31. Michelson, who had no use for any of the large mine owners, remarked: "Which shows that even Plunder Barons may be maligned, and that there is honor among them, when it is expedient to be honorable."

35. Berlin, *Silver Platter*, 184–91, 168; Smith, "Manuscripts and notes," carton 2a, "John W. Mackay, Bonanza King," 298–99.

CHAPTER 4. *Competitors*

1. *New York Tribune*, July 21, 1902; Don C. Seitz, *The James Gordon Bennetts, Father and Son: Proprietors of the New York Herald* (Indianapolis: Bobbs-Merrill Co., 1928), 49, 51, 55–66, 89.

2. Doten, *Journals*, vol. 2, 1055; Berlin, *Silver Platter*, 179–80.

3. Goodwin, *As I Remember Them*, 162.

4. Book of Deeds, Storey County Records, Nevada, book 27, May 7, 1870, 163; Fair and Sharon sell the Kirpatrick mill on April 1, 1880, book 44, 637–38.

5. Lord, *Comstock Mining and Miners*, 234; Angel, *History of Nevada*, 506.

6. Adolph Sutro, in *Alta California* (later the *Daily Alta*), April 20, 1860.

7. Lord, *Comstock Mining and Miners*, 234.

8. Smith, *History of the Comstock Lode*, 109.

9. Gustavus Myers, *History of the Great American Fortunes* (1907; reprint, New York: Random House, 1964), 153–54; Charles R. Morris, *The Tycoons: How Andrew Carnegie, John D. Rockefeller, Jay Gould, and J. P. Morgan Invented the American Supereconomy* (New York: Henry Holt & Co., 2005), 61.

10. *Gold Hill News*, September 22, 1869.

11. *Territorial Enterprise*, November 5, 1872; Lord, *Comstock Mining and Miners*, 237–42. See also Van Tilberg Clark's commentary in Doten, *Journals*, vol. 2, 1221–22.

12. Adolph Sutro, "Closing Argument of Adolph Sutro," April 22, 1872, 17, The Bancroft Library, University of California, Berkeley.

13. *Gold Hill News*, October 7, 1869; Angel, *History of Nevada*, 280.

14. Smith, *History of the Comstock Lode*, 104.

15. Sam Davis, *History of Nevada*, vol. 2, 1063.

16. Smith, "Manuscripts and notes," carton 2, notebook 4, 344, 357.

17. Lord, *Comstock Mining and Miners*, 303–4; *San Francisco Chronicle*, December 30, 1894; Smith, *History of the Comstock Lode*, 263–64.

18. Doten, *Journals*, vol. 2, 1040, 1230–31, 1250; Smith, *History of the Comstock Lode*, 148.

19. Sam Davis, *History of Nevada*, vol. 1, 425–26.

20. Smith, "Manuscripts and notes," carton 2a, "John W. Mackay, Bonanza King," 558.

21. Leslie, *California: A Pleasure Trip*, 280–81; *Territorial Enterprise*, March 28, April 19, 1878; Smith, *History of the Comstock Lode*, 152.

22. Alfred E. Davis, in *San Francisco Chronicle*, December 30, 1894.

23. *San Francisco Examiner*, December 30, 1894; Smith, *History of the Comstock Lode*, 145; Steve Gillis, *Memories of Mark Twain and Steve Gillis* (Sonora, Calif.: Banner, n.d.), 123–25; Drury, *An Editor*, 65; Smith, "Manuscripts and notes," carton 2a, "John W. Mackay, Bonanza King," 62; Goodwin, *As I Remember Them*, 180.

24. *San Francisco Examiner*, December 30, 1894; Wells Drury, in Smith, "Manuscripts and notes," carton 2, "Comstock Notes," section M, n.p.

25. Lord, *Comstock Mining and Miners*, 381.

26. Gillis, *Memories of Mark Twain*, 123.

27. *San Francisco Examiner*, December 30, 1894, July 21, 1902; Lingenfelter, *Hardrock Miners*, 61.

28. Smith, "Manuscripts and notes," carton 2a, "John W. Mackay, Bonanza King," 549; Lingenfelter, *Hardrock Miners*, 6–7; Ronald James, *Roar and the Silence*, 122.

29. *San Francisco Call*, July 21, 1902; H. H. Bancroft, *Chronicle of the Builders* (San Francisco: History Co., [1890]), 227–28.

30. Smith, "Manuscripts and notes," carton 2a, "John W. Mackay, Bonanza King," 552–53.

31. Goodwin, *As I Remember Them*, 180–81. For a different version of the same story, see *San Francisco Examiner*, December 30, 1894.

CHAPTER 5. *The Con. Virginia*

1. Rohrbough, "Mining and the Nineteenth-century West," in Deverell, *Companion*, 118–19.

2. For discussions of property, enterprise, and cities in the West, see Limerick, *Legacy of Conquest*, 69–73, 124–33; and Rodman W. Paul, *The Far West and the Great Plains in Transition, 1859–1900* (New York: Harper & Row, 1988), 92–104. For a discussion of Virginia City's relation to San Francisco, see Earl Pomeroy, *The Pacific Slope: A History of California, Oregon, Washington, Idaho, Utah, and Nevada* (New York: Alfred A. Knopf, 1966), 125.

3. J. Ross Browne, *Peep at Washoe and Washoe Revisited,* 179. Eliot Lord gave his perception in 1879 and 1880: "Viewed from the mountain summit above the line of the shafts the dingy heaps of rock and sand near the mouth of every pit and tunnel appeared like ant-hills rising imperceptibly from day to day." Lord, *Comstock Mining and Miners,* 221.

4. Grace Dangberg, *Conflict on the Carson: A Study of Water Litigation in Western Nevada* (Minden, Nev.: Carson Valley Historical Society, 1975), 17, 26–27; Townley, "Reclamation in Nevada," 211.

5. Sam Davis, *History of Nevada,* vol. 1, 406; Shinn, *Story of the Mine,* 100; DeQuille, *Big Bonanza,* 168.

6. Lord, *Comstock Miners and Mining,* 262.

7. Ibid., 260–62.

8. Lewis, *Silver Kings,* 107.

9. Rossiter W. Raymond, *Silver and Gold: An Account of the Mining and Metallurgical Industry of the United States* (1873; reprint, University of Michigan: Scholarly Publishing Office, n.d.) 141, 162–67.

10. John Debo Galloway, *Early Engineering Works Contributory to the Comstock, University of Nevada Bulletin,* vol. 41, no. 5 (Reno: Nevada State Bureau of Mines and the Mackay School of Mines, June 1947), 63; Dangberg, *Conflict on the Carson,* 10–11; Angel, *History of Nevada,* 124, 300–301; Smith, *History of the Comstock Lode,* 222–27.

11. Gray Brechin, *Imperial San Francisco: Urban Power, Earthly Ruin* (Berkeley: University of California Press, 1999), 78–79.

12. Galloway, *Early Engineering Works,* 67–74; DeQuille, *Big Bonanza,* 170–73; George Wharton James, *The Lake of the Sky: Lake Tahoe in the High Sierras of California and Nevada,* (1915; 2d ed., Pasadena: Radiant Life Press, 1921), 260; Angel, *History of Nevada,* 600–602.

13. Smith, *History of the Comstock Lode,* 119, 145, 148; Berlin, *Silver Platter,* 201; Henry R. Whitehill, *Biennial Report of the State Mineralogist of the State of Nevada for the Year 1871 and 1872* (Carson City: John J. Hill, state printer, 1872), 115; "Nevada Mining Companies collection, 1862–1901," letterbook #3, Hale & Norcross, March 1872–July 1873, 12, 41, The Huntington Library, 77–78; Lord, *Comstock Mining and Miners,* 309; Raymond, *Silver and Gold,* 1873, 146.

14. Lord, *Comstock Mining and Miners,* 41–49.

15. Smith, *History of the Comstock Lode,* 81, 85, 145–47; Henry DeGroot, in John J. Powell, *Nevada: The Land of Silver* (San Francisco: Bacon & Co., 1876), 73–75; Lord, *Comstock Mining and Miners,* 308–309; Raymond, *Silver and Gold,* 520–21.

16. Ferdinand Bacon Richtofen, *The Comstock Lode: Its Character, and the Probable Mode of Its Continuance in Depth* (San Francisco: Towne & Bacon, 1866), 73. Richthofen's is an exhaustive report, which investigated the geology of the Washoe range and the structure and nature of the vein; William Ashburner to Samuel Bowles, November 1865, as published in Samuel Bowles, *Across the Continent: A*

Summer's Journey to the Rocky Mountains, the Mormons, and the Pacific States with Speaker Colfax (1866; reprint, Ann Arbor: University Microfilms, 1966), 449.

17. Raymond, *Silver and Gold*, 1873, 146–47; Smith, *History of the Comstock Lode*, 148; DeGroot, in Powell, *Nevada: The Land of Silver*, 75–78.

18. Lord, *Comstock Mining and Miners*, 46, 309; Smith, *History of the Comstock Lode*, 148. The former owner of a large part of what became the Consolidated Virginia, a man named Richard Sides, had an intriguing history in Nevada. He had come with two partners to Carson Valley in 1854 and claimed land for his Clear Creek ranch. A leader in the territory's secessionist movement against the government of Utah, he was involved in the vigilante action that resulted in the hanging of powerful settler William "Lucky Bill" Thorington at the Clear Creek ranch. When Brigham Young recalled the Mormons from Nevada to Salt Lake City, Sides and a partner named Jacob Rose took possession of a sawmill in Washoe Valley owned by Mormon elder Orson Hyde. A refusal to pay Hyde the twenty thousand dollars he believed the property to be worth prompted the issuance of "Hyde's curse" from the elder at Salt Lake City. The curse read in part, "The said R. D. Sides and Jacob Rose shall be living and dying advertisements of God's displeasure, in their persons, in their families, in their substances; and this demand of ours remaining uncancelled, shall be to the people of Carson and Washoe Valleys as was the ark of God among the Philistines. (See 1st Sam. Fifth chapter.) You shall be visited of the Lord of Hosts with thunder and with earthquakes and with floods, with pestilence and with famine until your names are not known amongst men." Years later a slide covered the mill site with sand, rendering it practically worthless. But another affliction struck Sides with more immediacy. Nearly half of the 1,010 feet of what became the Con. Virginia, 500 feet, was the Dick Sides claim, which in previous years of work had produced not a ton of ore. After he sold it, its lower depths proved to be the "richest spot in the world," containing the "Big Bonanza" (Angel, *History of Nevada*, 31; Sam Davis, *History of Nevada*, vol. 1, 232; Smith, *History of the Comstock Lode*, 146; DeQuille, *Big Bonanza*, 363, 371).

19. Smith, *History of the Comstock Lode*, 148–49; DeGroot, in Powell, *Nevada: The Land of Silver*, 79.

20. *Gold Hill News*, May 1 and May 25, 1872. The richness of the ore gave the strike the title the "Great Bonanza," a name later usurped by the Con. Virginia.

21. Smith, *History of the Comstock Lode*, 131, 148–49.

22. DeQuille, *Big Bonanza*, 405–6; Smith, *History of the Comstock Lode*, 87.

23. See, for example, *Gold Hill News*, May 4, May 11, May 25, July 27, August 10, September 21, 1872.

24. DeQuille, *Big Bonanza*, 361; John Russell Young, *Men and Memories*, 444.

25. *Gold Hill News*, May 11, 1872; Lingenfelter, *Hardrock Miners*, 17–18. In 1880 metallurgist and mining expert Frederick Roeser noted that while the face of the head of the sledge should be flat, it should be beveled down until the diameter is reduced by nearly one-half. This causes the sledge to fly off from the head of the

drill in case of a false blow. He comments that "this requirement however, is seldom provided for." Roeser (Frederick) Collection, notebook #2, "Mining Machinery, 6-7, The Huntington Library.

26. Max Crowell, in collaboration with Nevada State Writers' Project Work Projects Administration, *A Technical Review of Early Comstock Mining Methods* (Reno: Nevada State Bureau of Mines, 1941), 6; DeGroot, in Powell, *Nevada: The Land of Silver*, 80; Smith, *History of the Comstock Lode*, 151n.

27. *Gold Hill News*, July 27, August 3, August 17, September 19, 1872; Smith, *History of the Comstock Lode*, 150.

28. Smith, *History of the Comstock Lode*, 149-50, 151n; *Gold Hill News*, September 14, 1872.

29. Smith, *History of the Comstock Lode*, 157-58; *Territorial Enterprise*, September 19, 1872; *Gold Hill News*, November 30, December 7, 1872.

30. *Territorial Enterprise*, September 19, 1872; Berlin, *Silver Platter*, 199-202; Smith, *History of the Comstock Lode*, 149-51, 149n.

31. Berlin, *Silver Platter*, 201.

32. *Gold Hill News*, February 1, February 8, 1873.

33. Lord, *Comstock Mining and Miners*, 310; Smith, *History of the Comstock Lode*, 151; *Gold Hill News*, November 23, 1872.

34. Lord, *Comstock Mining and Miners*, 310; *Territorial Enterprise*, March 20, 1873; Smith, *History of the Comstock Lode*, 151; *Gold Hill News*, March 22, 1873.

35. *Territorial Enterprise*, March 20, 1873. The Curtis–DeQuille friendship is mentioned by DeQuille in a letter in Richard A. Dwyer and Richard E. Lingenfelter, *Dan DeQuille the Washoe Giant: A Biography and Anthology* (Reno: University of Nevada Press, 1990), 29. The colorful Curtis went on to work for the Ophir, the Justice, and other mines, becoming involved in two scandals in later years. He embezzled funds to build a boardinghouse, run by his mistress and kept at full occupancy, by demanding that miners working for him board there. While at the Justice mine, he had a row with the superintendent of the Alta over a vein on the boundary of two. They each hired gunmen to prove their assertions. The affray was only halted by the intercession of several hundred men from the miners' union, who marched to the site to keep the peace. Later arbitration involving other mine superintendents, led by Fair, resolved the dispute. *Gold Hill News*, January 15 and January 17, 1878; Lingenfelter, *Hardrock Miners*, 28, 56-57.

36. *Gold Hill News*, March 29, 1873; DeQuille, *Big Bonanza*, 362, 365.

37. *Gold Hill News*, March 29, April 12, 1873; DeGroot, in Powell, *Nevada: The Land of Silver*, 80-81; Crowell, *Technical Review*, 5.

38. Berlin, *Silver Platter*, 203-7; Smith, "Manuscripts and notes," carton 2a, "John W. Mackay, Bonanza King," 311.

39. Galloway, *Early Engineering Works*, 67-74; DeQuille, *Big Bonanza*, 170-73; Henry R. Whitehill, *Biennial Report of the State Mineralogist of the State of Nevada for the Year 1873 and 1874* (Carson City: John J. Hill, state printer, 1874), 90. The system continues to supply water to Virginia City to the present day.

40. DeGroot, in Powell, *Nevada: The Land of Silver*, 81.

41. Ibid., 81; DeQuille, *Big Bonanza*, 365.

42. Smith, *History of the Comstock Lode*, 153; New York *World*, February 27, 1902; Smith, "Manuscripts and notes," carton 2, "Comstock Notes," 4, and section R, n.p.

43. Smith, *History of the Comstock Lode*, 153–54; "Mining and Scientific Press," January 16, 1875, in Smith, *History of the Comstock Lode*, 153.

44. Smith, *History of the Comstock Lode*, 154–55; Drury, *An Editor*, 211; DeQuille, *Big Bonanza*, 365; Berlin, *Silver Platter*, 207.

45. Smith, *History of the Comstock Lode*, 155; T. A. Hopkins, "Report on The Consolidated Virginia Mine and The Underground Mill," August 9, 1926, 3, DeLamare Library, University of Nevada, Reno; DeGroot, in Powell, *Nevada: The Land of Silver*, 85.

CHAPTER 6. *Bonanza Silver and Kings*

1. Smith, "Manuscripts and notes," carton 2a, "John W. Mackay, Bonanza King," 307; Berlin, *Silver Platter*, 208–9; Harriet Land Levy, in Lewis, *This Was San Francisco: Being First-Hand Accounts of the Evolution of One of America's Favorite Cities*, ed. Oscar Lewis (New York: David McKay Co., 1962), 223.

2. Berlin, *Silver Platter*, 208–11; Doten, *Journals*, vol. 2, 1220–24; Kathryn Dunn Totton, "They Are Doing So to a Liberal Extent Here Now," in James and Raymond, *Comstock Women*, 68–69; Lowe, "Opium in Comstock Society," in James and Raymond, *Comstock Women*, 109–10. Opium smoking was not outlawed until two years later, when its popularity among middle-class youths was publicly revealed (104).

3. Will Irwin, "The City That Was," *New York Sun*, April 21, 1906; Levy, in Lewis, *This Was San Francisco*, 224; Neville, *Fantastic City*, 2.

4. Smith, "Manuscripts and notes," carton 2, "Mackay. W Nev. Bank Postal Co, Cable Cos. Posted 6," section s, n.p.; Berlin, *Silver Platter*, 448.

5. Berlin, *Silver Platter*, 213, 235.

6. Zoeth Skinner Eldredge, *History of California* (New York: Century History Co., 1915) 434–35.

7. "Commercial Herald & Market Review," January 12, 1872, in Tilton, *William Chapman Ralston*, 161–62; Neville, *Fantastic City*, 198.

8. DeGroot, in Powell, *Nevada: The Land of Silver*, 83–84; *New York Tribune*, August 27, 1875; Angel, *History of Nevada*, 126.

9. *Virginia Chronicle* clipping, quoted in Doten, *Journals*, vol. 2, 1218.

10. Smith, *History of the Comstock Lode*, 11–12; Shinn, *Story of the Mine*, 146.

11. DeGroot, in Powell, *Nevada: The Land of Silver*, 84–85; Lingenfelter, *Hardrock Miners*, 18; Raymond, *Silver and Gold*, 1873, 488.

12. DeGroot, in Powell, *Nevada: The Land of Silver*, 84–85; Smith, *History of the Comstock Lode*, 158–59; DeQuille, *Big Bonanza*, 254–61, 365.

13. Marye, *From '49 to '83,* 167–68.

14. DeGroot, in Powell, *Nevada: The Land of Silver,* 86; *San Francisco Chronicle,* January 10, 1875.

15. DeGroot, in Powell, *Nevada: The Land of Silver,* 86; John Russell Young, *Men and Memories,* 442; DeQuille, *Big Bonanza,* 376.

16. DeQuille, *Big Bonanza,* 371–72; Whitehill, *Biennial Report . . . 1873 and 1874,* 90.

17. Marye, *From '49 to '83,* 168–69.

18. *San Francisco Chronicle,* January 4, 1875.

19. Stewart biographer Russell Elliot commented: "[To Stewart] what mattered was not philosophical or legal right or wrong, but only who was victorious." Russell R. Elliot, *Servant of Power: A Political Biography of Senator William M. Stewart,* (Reno: University of Nevada Press, 1983), 18–20, 26–32, 45. The quote on matching any adversary was taken by Elliot from Lord, *Comstock Mining and Miners,* 146. Elliot comments that Lord interviewed Stewart a number of times and "considered Stewart an exceptional person" (289n.).

20. Elliot, *Servant of Power,* 23–24, 32.

21. Angel, *History of Nevada,* 122–23; Smith, *History of the Comstock Lode,* 205.

22. Angel, *History of Nevada,* 124.

23. *San Francisco Chronicle,* November 14, 1885.

24. Angel, *History of Nevada,* 124–26; Smith, *History of the Comstock Lode,* 205.

25. Angel, *History of Nevada,* 123, 126; Smith, *History of the Comstock Lode,* 205, 260–61. Gross bullion produced in Storey County in 1866 was reported by Angel to be $11,951,876. The gross production for the state that year was $14,907,895, rising to $40,784,469 in 1875. (Sam Davis, *History of Nevada,* vol. 1, 352–55.)

26. Sam Davis, *History of Nevada,* vol. 1, 419; Angel, *History of Nevada,* 126–30.

27. Effie Mona Mack, *Nevada: A History of the State from the Earliest Times through the Civil War* (Glendale, Calif.: Arthur H. Clark Co., 1936), 442–43; Smith, *History of the Comstock Lode,* 159–60; *Nevada State Journal,* June 24, July 1, 1875.

28. *Nevada State Journal,* April 18, 1875; Sam Davis, *History of Nevada,* vol. 1, 419.

29. Tilton, *William Chapman Ralston,* 338; Ira B. Cross, *Financing an Empire: The History of Banking in California* (Chicago: S. J. Clarke Publishing Co., 1927), vol. 1, 412. McLane had been in California since 1846. He was present at the raising of the U.S. flag at Monterey and was one of Fremont's representatives at the treaty of Chauenga. In 1868, when he was appointed president of Wells Fargo, he moved to New York. Upon his return to California in 1875, the Bank of California wanted him to again associate himself with it, but he allied himself with the Nevada Bank of San Francisco instead.

CHAPTER 7. *1875*

1. Matthew Josephson, *The Robber Barons: The Great American Capitalists 1861–1901* (1934; reprint, New York: Harcourt, Brace & World, 1962), 92–93, 164–65;

Kevin Phillips, *Wealth and Democracy: A Political History of the American Rich* (New York: Broadway Books, 2002), 304–5; Morris, *Tycoons,* 137–38.

2. Tilton, *William Chapman Ralston,* 324–29; Louis M. Hacker and Benjamin B. Kendrick, with the collaboration of Helene S. Zahler, *The United States since 1865* (1932; 4th ed., New York: Appleton-Century-Crofts, 1949), 176–82; Morison and Commager, *Growth of the American Republic,* 245; Smith, *History of the Comstock Lode,* 142–43; "The Big Bonanza," *New York Tribune,* August 27, 1875.

3. Smith, *History of the Comstock Lode,* 162–63; Sam Davis, *History of Nevada,* vol. 1, 421; *San Francisco Call,* November 14, 1885; C. B. Glasscock, *Lucky Baldwin: The Story of an Unconventional Success* (n.d.; reprint, Reno: Silver Syndicate Press, 1993), 142, 156–64; *San Francisco Chronicle,* January 9, 1875.

4. Smith, *History of the Comstock Lode,* 164–65, 171–72; Hinkle and Hinkle, *Sierra-Nevada Lakes,* 299; *San Francisco Chronicle,* January 9, 1875.

5. Smith, *History of the Comstock Lode,* 174–76, 179; Dwyer and Lingenfelter, *Dan Dequille,* 29; Lord, *Comstock Mining and Miners,* 316.

6. Smith, *History of the Comstock Lode,* 177; Robert Irving Washow, *Jay Gould: The Story of a Fortune* (New York: Greenberg, Publisher, 1928), 14, 137–38; Charles E. DeLong to Mrs. Elida DeLong, January 31, 1875, The Bancroft Library.

7. DeQuille, *Big Bonanza,* 368–70; *New York World,* July 27, 1902. Regarding investing, Mackay told Harry Gorham that if he, Mackay, had stayed out of speculating on stocks, he would have been ten million dollars better off. Harry M. Gorham, *My Memories of the Comstock* (Gold Hill, Nev.: Gold Hill Publishing, 2005), 22.

8. See Wm. Ralston to Chas. E. De Long, n.p., n.d., "Ralston Correspondence," The Bancroft Library.

9. Cross, *Financing an Empire,* vol. 1, 402; Asbury Harpending, *The Great Diamond Hoax: And Other Stirring Incidents in the Life of Asbury Harpending,* ed. James H. Wilkins (San Francisco: James H. Barry, 1913), 274; B. E. Lloyd, *Lights and Shades in San Francisco* (1876; facsimile ed., Berkeley: Berkeley Hills Books, 1999), 333; Smith, *History of the Comstock Lode,* 189.

10. *Nevada State Journal,* May 28, 1875; "The Big Bonanza," *New York Tribune,* August 27, 1875.

11. Tilton, *William Chapman Ralston,* 326–28, 354; Cross, *Financing an Empire,* vol. 1, 398; "Largest Hotel in the United States Now Going Up Here," San Francisco Real Estate Circular, December, 1873, online Museum of the City of San Francisco; http://www.sfmuseum.org/hist1/1873.html (accessed October 31, 2001); David Lavender, *Nothing Seemed Impossible: William C. Ralston and Early San Francisco* (1975; reprint, Palo Alto, Calif.: American West Publishing Co., 1981), 371; Eldredge, *History of California,* 440–41; Dangberg, *Conflict on the Carson,* 21–22.

12. Smith, *History of the Comstock Lode,* 179, 181.

13. *San Francisco Chronicle,* August 28, 1875; George Lyttleton Upshur, *As I Recall Them: Memories of Crowded Years* (New York: Wilson-Erickson, 1936), 92–93, 95. This incident gave rise to the widespread story that at the bank's collapse, O'Brien attempted to place a demijohn of whiskey and several glasses on the bank

counter but was prevented from doing so by Mackay. (See Drury, *An Editor*, 119, and Sam Davis, *History of Nevada*, vol. 1, 418.) The story is refuted by Berlin, *Silver Platter*, 215.

14. Eldredge, *History of California*, 438; Smith, *History of the Comstock Lode*, 189.

15. *San Francisco Call*, August 28, 1875; Julian Dana, *The Man Who Built San Francisco: A Study of Ralston's Journey with Banners* (New York: Macmillan Co., 1936), 348; Cross, *Financing an Empire*, vol. 1, 403–4.

16. Tilton, *William Chapman Ralston*, 339–41; Cross, *Financing an Empire*, vol. 1, 399.

17. Smith, *History of the Comstock Lode*, 179n; Sam Davis, *History of Nevada*, vol. 1, 631; Angel, *History of Nevada*, 596; *Gold Hill News*, August 27, 1875.

18. *Alta*, August 28, 1875; *Gold Hill News*, August 28, 1875.

19. H. L. Slosson, in Smith, "Manuscripts and notes," carton 2a, "John W. Mackay, Bonanza King," 449, 549. That evening Sharon repeated the sentiment about Ralston to another listener. As Ralston's friends crowded into a parlor where the body lay in a temporary coffin, he turned to bank secretary Stephen Franklin and said: "Best thing he could have done." Gertrude Atherton, *California: An Intimate History* (1914; rev. ed, New York: Horace Liveright, 1927), 279.

20. Cross, *Financing an Empire*, vol. 1, 399–401.

21. *San Francisco Chronicle*, August 27–29, 1875; Ralston biographer Cecil Tilton commented that hundreds of citizens were "incensed" at the Bonanza Kings and that their bank "suffered severely for some time as a result of these people's feelings." But Tilton also said that there appeared to be no malicious competition between Ralston and Flood, commenting further: "Actually also there appears little personal enmity. Flood was a great person, strong, honest. His cause was his own. So was Ralston's cause his own." *William Chapman Ralston*, 353–54, 406.

22. *Alta*, August 27, 1875.

23. *San Francisco Chronicle*, August 30, 1875.

24. Doten, *Journals*, vol. 2, 1255–56.

25. Ibid., 1258–61; "The Big Bonanza," *New York Tribune*, August 27, 1875.

26. *San Francisco Chronicle*, October 28, 1875; Smith, *History of the Comstock Lode*, 192; *Gold Hill News*, October 27, 1875; *Territorial Enterprise*, October 27, 1875.

27. Goodwin, *As I Remember Them*, 161. Goodwin originally related this anecdote in the *San Francisco Call*, July 21, 1902; Smith, *History of the Comstock Lode*, 193; *San Francisco Chronicle*, October 28, 1875.

28. Smith, *History of the Comstock Lode*, 192; *Territorial Enterprise*, October 28, 1875; S. T. Curtis, in Rossiter W. Raymond, *Statistics of Mines and Mining in the States and Territories West of the Rocky Mountains* (Washington, D.C.: Government Printing Office, 1877), 148; *San Francisco Chronicle*, October 28, 1875.

29. DeGroot, in Powell, *Nevada: The Land of Silver*, 93; Smith, *History of the Comstock Lode*, 193; Lord, *Comstock Mining and Miners*, 327.

30. Smith, *History of the Comstock Lode*, 193; *Gold Hill News*, October 27, 1875;

Territorial Enterprise, October 28, 1875; Smith, "Manuscripts and notes," carton 2, "Mackay. W Nev. Bank Postal(?) Co, Cable Cos. Posted 6," section S, n.p.; *San Francisco Chronicle,* October 28, 1875.

31. *San Francisco Chronicle,* October 28, October 31, 1875; Raymond, *Statistics of Mines and Mining,* 148, 157; Angel, *History of Nevada,* 325; *Carson Valley News,* November 6, 1875; Jake Highton, *Nevada Newspaper Days: A History of Journalism in the Silver State* (Stockton, Calif.: Heritage West Books, 1990), 74–75; Doten, *Journals,* vol. 2, 1262–63; *San Francisco Chronicle,* October 31, 1875.

32. *San Francisco Chronicle,* October 27, 1875.

33. *Ibid.,* October 28, 1875.

34. Smith, *History of the Comstock Lode,* 194.

35. *San Francisco Chronicle,* October 28, 1875; *Nevada State Journal,* January 25, 1876.

36. *San Francisco Chronicle,* October 31, 1875.

37. Smith, *History of the Comstock Lode,* 39–40; Rev. T. Tubman, in Sam Davis, *History of Nevada,* vol. 1, 544; Ronald James, *Roar and the Silence,* 201. See also Lyman, *Saga of the Comstock Lode,* 86–87, 241–42.

38. Grant Smith speculated: "This reticence of Mr. Mackay was doubtless out of consideration for his wife and his mother-in-law, and the fact that his boys were being brought up in the Catholic faith." (Smith, "Manuscripts and notes," carton 2a, "John W. Mackay, Bonanza King," 7–8, 316; Goodwin, *As I Remember Them,* 161.) In an obituary in the *San Francisco Call,* July 21, 1902, Goodwin said Fr. Manogue drew $450,000 on Mackay's personal account, later adjusting the figure for his book.

39. Raymond, *Statistics of Mines and Mining,* 148. The Ophir did raise ore from its upper levels a month after the fire, but while it was rebuilding its hoisting infrastructure and works above ground, its 1,500-foot level filled with water to a depth of 16 feet and had to be drained before meaningful mining was again pursued.

40. Lord, *Comstock Mining and Miners,* 328; Goodwin, *As I Remember Them,* 160–61. Some years later Mackay contributed ten thousand dollars to a famine relief fund for Ireland. (Smith, "Manuscripts and notes," carton 2a, "John W. Mackay, Bonanza King," 556.)

41. Lord, *Comstock Mining and Miners,* 328; DeQuille, *Big Bonanza,* 435–36; Smith, "Manuscripts and notes," carton 2, notebook 2, 2.

42. Raymond, *Statistics of Mines and Mining,* 154–55.

43. Ibid., 155–57.

44. *Annual Report of the Consolidated Virginia Mining Company for 1875* (San Francisco: Office of the Daily Stock Exchange, 1876), 8–9; *San Francisco Bulletin,* August 5, 1878; Smith, *History of the Comstock Lode,* 194–95.

CHAPTER 8. *Responses*

1. H.J. Ramsdell, "The Great Nevada Flume," in Williams, *Pacific Tourist,* 219–23.

2. Lewis, *Silver Kings*, 161–62.

3. John Taylor Waldorf, *A Kid on the Comstock: Reminiscences of a Virginia City Childhood*, edited, with introduction and commentary, by Dolores Waldorf Bryant (1968; reprint, Vintage West Series, Reno: University of Nevada Press, 1991), 62–63; Smith, *History of the Comstock Lode*, 266.

4. Ronald James, *Roar and the Silence*, 206–7.

5. *San Francisco Examiner*, December 30, 1894; Lewis, *Silver Kings*, 119; Goodwin, *As I Remember Them*, 183–84.

6. Smith, "Manuscripts and notes," carton 2, "Comstock Notes," section S, n.p.; H. L. Slossen, in Smith, "Manuscripts and notes," carton 2a, "John W. Mackay, Bonanza King," 550–51.

7. Lewis, *Silver Kings*, 154–57; Leslie, *California: A Pleasure Trip*, 280–81; Smith, *History of the Comstock Lode*, 145.

8. Lina Fergusson Browne, *J. Ross Browne*, 405.

9. Smith, *History of the Comstock Lode*, 199.

10. Berlin, *Silver Platter*, 218; *San Francisco Chronicle*, reprinted in *Nevada State Journal*, January 25, 1876.

11. Lewis, *Silver Kings*, 162–63; Smith, *History of the Comstock Lode*, 199.

12. Smith, *History of the Comstock Lode*, 199–201.

13. Sam Davis, *History of Nevada*, vol. 2, 724–25; Smith, *History of the Comstock Lode*, 197. Alf Doten observed in 1883 that Mrs. Bowers "picks up plenty of pocket money" as a spiritualist. "Some of her predictions have been literally, squarely (and unavoidably) verified, but numerous others have not, by a long chalk, and never can be, but then she is very hard of hearing, not being able to distinguish the jingle of a quarter from the ringing of a church bell for the last thirty years" (*Journals*, vol. 2, 1447).

14. The Combination Shaft was eventually sunk well below the three-thousand-foot level and was the last of the Comstock's deep-mining shafts, closing on October 16, 1886. Smith, *History of the Comstock Lode*, 90.

15. *Daily Stock Report*, April 14, 15, 1876; Smith, *History of the Comstock Lode*, 90, 269.

16. Smith, *History of the Comstock Lode*, 177, 200–201; *Nevada State Journal*, April 1, 1876; *Daily Stock Report*, April 14, 15, 1876.

17. Ronald James, *Roar and the Silence*, 130, 207–8. James says that Van Bokkelen's pet monkey may have detonated the explosion that killed him and lists a number of individuals killed in the mishap. He mentions that many were wounded, reporting further that "rescue workers found no trace of the monkey."

18. Ronald James, *Roar and the Silence*, 199.

19. Berlin, *Silver Platter*, 218–21; Lewis, *Silver Kings*, 77; Lewis, *This Was San Francisco*, 207–8; Gertrude Atherton, *My San Francisco: A Wayward Biography* (New York: Bobbs-Merrill Co., 1946), 32–33; *Nevada State Journal*, January 25, 1876.

20. Sven Beckert, *The Monied Metropolis: New York City and the Consolidation of the American Bourgeoisie, 1850–1896* (Cambridge: Cambridge University Press, 2001), 3–10.

21. Berlin, *Silver Platter*, 230–33.

22. Samuel Ward McAllister, "A Glimpse of High Society," in *Empire City: New York Through the Centuries*, ed. Kenneth T. Jackson and David S. Dunbar (New York: Columbia University Press, 2002), 356–58; Arthur Meir Schlesinger, *The Rise of the City, 1878–1898*, vol. 10, *A History of American Life* (1933; reprint, New York: Macmillan Co., 1969), 152–53.

23. Beckert, *Monied Metropolis*, 209, 211.

24. Berlin, *Silver Platter*, 218, 223–25.

25. Ibid., 226–35, 247–48.

26. Julia Cooley Altrocchi, *The Spectacular San Franciscans* (New York: E. P. Dutton & Co., 1949), 271.

27. *Nevada State Journal*, May 17, September 29, 1876. One of the attractions was the twenty-thousand-dollar silver works (the money allocated by the Nevada Legislature) built in the southwest corner of the 285-acre fairgrounds. It was alongside the Machinery Building, behind the massive Total Abstinence Fountain, which featured five towering sculptured figures, the central one of whom was Moses. Four mines, the Con. Virginia, the California, the Ophir, and the Belcher, sent quartz that kept the works running throughout the six months of the exhibition. J. S. Ingram, *Centennial Exposition Described and Illustrated* (Philadelphia: Hubbard Bros., 1876), 193–94, http://fax.libs.uga.edu (accessed February 1, 2006). The Total Abstinence Fountain still stands in Philadelphia's West Fairmount Park.

28. Smith, *History of the Comstock Lode*, 200; Smith, "Manuscripts and notes," carton 2a, "John W. Mackay, Bonanza King," 310; Berlin, *Silver Platter*, 240–44.

29. *The Stock Exchange*, August 1, 1876, in Smith, "Manuscripts and notes," carton 2a, "John W. Mackay, Bonanza King," 467.

30. *Annual Report . . . 1875*, 9; Smith, *History of the Comstock Lode*, 195, 200.

31. John P. Young, *Journalism in California* (San Francisco: Chronicle Publishing Co., 1915), 79–83; *San Francisco Chronicle*, January 10, 1875. For its opinion of the Bank Ring's practices, see "The Stock-Jobbing Juggernaut," *San Francisco Chronicle*, May 20, 1875. Grant Smith, in *History of the Comstock Lode*, commented that both Dewey and the *Chronicle* managers had lost money in Con. Virginia "and both proposed that the Bonanza Firm should make restitution" (218).

32. Smith, *History of the Comstock Lode*, 200–202; *Nevada State Journal*, April 1, 1876; Lord, *Comstock Mining and Miners*, appendix, table V, 430–34.

33. Smith, "Manuscripts and notes," carton 2, "Mackay duplicates and discards 8," n.p.

34. Marye, *From '49 to '83*, 79; Smith, *History of the Comstock Lode*, 219.

35. The quote is from an article titled "Who is Squire P. Dewey" in an undated publication from the *San Francisco News Letter*, in Grant Horace Smith's papers,

"History of the Comstock Lode," Special Collections, Getchell Library, University of Nevada, Reno.

36. *San Francisco Chronicle,* January 12, 1877; *Mining and Scientific Press,* January 20, 1877; S. P. Dewey, "The Bonanza Mines of Nevada: Gross Frauds in the Management Exposed," 1878, Special Collections, Getchell Library, University of Nevada, Reno.

37. Smith, "Manuscripts and notes," carton 2a, "John W. Mackay, Bonanza King," 313 and 549; *New York Times,* August 23, 1890.

38. *San Francisco Chronicle,* January 12, 1877; *San Francisco Post,* January 12, 1877; *San Francisco Mail,* January 12, 1877; *San Francisco News Letter,* January 12, 1877; "Who is Squire P. Dewey."

39. Lord, *Comstock Mining and Miners,* appendix, table V, 433–35.

40. Smith, *History of the Comstock Lode,* 204.

CHAPTER 9. *Status and Scandal*

1. Beckert, *Monied Metopolis,* 156, 262; Lewis, *Silver Kings,* 90.

2. Berlin, *Silver Platter,* 242–43 and 251; Beckert, *Monied Metropolis,* 260; *New York Tribune,* February 22, 1885.

3. *Territorial Enterprise,* April 6, 1881.

4. *Nevada State Journal,* June 10, September 3, 1880, July 15, 1884; Lewis, *Silver Kings,* 104; Berlin, *Silver Platter,* 247.

5. Berlin, *Silver Platter,* 250; Smith, "Manuscripts and notes," carton 2a, "John W. Mackay, Bonanza King," 663.

6. Smith, "Manuscripts and notes," carton 2a, "John W. Mackay, Bonanza King," 316 and 663; Lewis, *Silver Kings,* 88.

7. Berlin, *Silver Platter,* 252–57; William Ralston Balch, in Lewis, *Silver Kings,* 88–89; *Nevada State Journal,* February 27, 1880.

8. Lewis, *Silver Kings,* 92–93; Smith, "Manuscripts and notes," carton 2, notebook 1b, 311; Smith, "Manuscripts and notes," carton 2a, "John W. Mackay, Bonanza King," 663.

9. Smith, "Manuscripts and notes," carton 2a, "John W. Mackay, Bonanza King," 312–13; Lewis, *Silver Kings,* 85; Smith, "Manuscripts and notes," carton 2, notebook 1b, 317.

10. Smith, "Manuscripts and notes," carton 2a, "John W. Mackay, Bonanza King," 309, 312–14.

11. Berlin, *Silver Platter,* 262–64; *Territorial Enterprise,* March 28, 1878; Ronald James, *Roar and the Silence,* 108.

12. *Territorial Enterprise,* June 12, 1878; Ronald James, *Roar and the Silence,* 207.

13. *Territorial Enterprise,* March 20, April 6, 1881, March 5, 1882; Berlin, *Silver Platter,* 292–94; *Nevada State Journal,* June 19, 1883.

14. Edmond M Gagey, *The San Francisco Stage: A History* (New York: Columbia

University Press, 1950), 117; John McCullough, *Chicago Times,* December 31, 1880, in Smith, "Manuscripts and notes," carton 2a, "John W. Mackay, Bonanza King," 561.

15. Lewis, *Silver Kings,* 54; Gagey, *San Francisco Stage,* 152; Smith, "Manuscripts and notes," carton 2a, "John W. Mackay, Bonanza King," 557; John McCullough, *Chicago Times,* December 31, 1880, in Smith, "Manuscripts and notes," carton 2a, "John W. Mackay, Bonanza King," 561; *San Francisco Examiner,* July 21, 1902.

16. Washow, *Jay Gould,* 125–27; New York *World,* November 3, 1901.

17. Josephine Mansfield abandoned Stokes once he was jailed. She married a wealthy lawyer who died in an insane asylum and lived into her nineties in Paris as a lady of means. Richard O'Connor, *Gould's Millions* (New York: Doubleday & Co., 1962), 119–21; Josephson, *Robber Barons,* 156–57; *New York Times,* November 28, 1871, January 6–9, 1872, November 3, 1901; New York *World,* November 3, 1901.

18. Smith, "Manuscripts and notes," carton 2, notebook v, n.p., "Comstock Notes," section R, n.p., and Smith, "Manuscripts and notes," carton 2a, "John W. Mackay, Bonanza King," 550; Chandos Fulton, in *New York Times,* July 27, 1902.

19. C. C. Goodwin, Eliot Lord, and E. C. Bradley, Vice President Postal Telegraph, in Smith, "Manuscripts and notes," carton 2, "Comstock Notes," section R, n.p.; Edward Chase Kirkland, *Dream and Thought in the Business Community, 1860–1900* (1956; reprint, Chicago; Ivan R. Dee, 1990), 37, 42.

20. *New York Times,* February 22, 1889; Joseph King, *History of the San Francisco Stock and Exchange Board, by the Chairman, Jos. L. King* (San Francisco: J. L. King, 1910), 212; Lewis, *Silver Kings,* 235.

21. Upshur, *As I Recall Them,* 104–5. The painting mentioned hung in Clarence Mackay's New York office in 1939, fifty years later.

22. Lewis, *Silver Kings,* 223–24, 265.

23. Smith, *History of the Comstock Lode,* 119; Atherton, *My San Francisco,* 35–36; Lewis, *Silver Kings,* 225–26.

24. Lewis, *Silver Kings,* 263–68; Berlin, *Silver Platter,* 266.

25. See Angel, *History of Nevada,* 126–30; *Territorial Enterprise,* October 5, 1880; Smith, *History of the Comstock Lode,* 204–6.

26. Robert Lewis Stevenson, in Pomeroy, *Pacific Slope,* 127.

27. Dewey, "Bonanza Mines of Nevada," 19–20; Smith, *History of the Comstock Lode,* 220–21.

28. Lord, *Comstock Mining and Miners,* 248; Smith, *History of the Comstock Lode,* 262.

29. Lord, *Comstock Mining and Miners,* appendix, table II, 419–21.

30. *Territorial Enterprise,* December 21, 1880; Dewey, "Bonanza Mines of Nevada," appendix, 60–66; "Burke v. J.C. Flood, Et Al," Superior Court of San Francisco, 1880, 2–3, Special Collections, Getchell Library, University of Nevada, Reno; Smith, "Dewey's Suits," Special Collections, Getchell Library, University of Nevada, Reno; Smith, *History of the Comstock Lode,* 218–22.

31. *San Francisco Chronicle,* December 27, 1878.

32. B. E. Lloyd, *Lights and Shades,* 405; *San Francisco Bulletin,* December 28, 1878.

33. *Territorial Enterprise,* December 29, 1878.

34. B. E. Lloyd, *Lights and Shades,* 409–11.

35. *San Francisco Chronicle,* December 27, 1878.

36. Lewis, *Silver Kings,* 159–60; Smith, "Manuscripts and notes," carton 2, "Comstock Notes," section Q, 2; *Carson Appeal,* September 28, 1883. The Appeal reported that upon her release Amelia Smallman spoke freely to friends that she would seek to avenge herself by shooting Mackay. "That she would shoot Mackay on sight is also not believed improbable by those who know her." Nothing came of the alleged threat.

CHAPTER 10. *To Try Fortune No More*

1. R. K. Colcord, in Smith, "Manuscripts and notes," carton 2, "Comstock Notes," 8.

2. Smith, "Manuscripts and notes," carton 2a, "John W. Mackay, Bonanza King," 467.

3. *Territorial Enterprise,* November 13, 1877; *Gold Hill News,* January 17, 1878.

4. Lewis, *Silver Kings,* 158; *Territorial Enterprise,* November 13, 1877; Doten, *Journals,* vol. 3, 2131. Fair's contention that Flood was not familiar with the workings of the mine was true. When Flood was examined in the Burke suit, he testified that he did not know which of their mills milled Con. Virginia ore, did not know whether milling contracts were made orally or in writing, could not say if the slimes and tailings were collected in ponds nor anything else about them, did not know where the crude bullion was sent for melting into bars, did not know who paid the expenses of the assay office. The *New York Times* of November 23, 1880, titled the article on his testimony "What Mr. Flood Does Not Know."

5. Lewis, *Silver Kings,* 120–21 and 162–63; Smith, *History of the Comstock Lode,* 224.

6. *New York Times,* July 18, 1885; Gorham, *My Memories of the Comstock,* 22–26; Doten, *Journals,* vol. 2, 1325.

7. Smith, *History of the Comstock Lode,* 222–24; *San Francisco Bulletin,* August 5, 26, 1878; *San Francisco Chronicle,* September 1, 1878.

8. Doten, *Journals,* vol. 2, 1329; King, *History of the San Francisco Stock and Exchange Board,* 155.

9. Marye, *From '49 to '83,* 197–99.

10. Smith, *History of the Comstock Lode,* 224–25; King, *History of the San Francisco Stock and Exchange Board,* 155.

11. Smith, *History of the Comstock Lode,* 225–26; Doten, *Journals,* vol. 2, 1334–35; Drury, *An Editor,* 233.

12. Doten, *Journals,* vol. 2, 1335; Lord, *Comstock Mining and Miners,* appendix, table V, 435.

13. *New York Times,* July 18, 1885.

14. Smith, *History of the Comstock Lode,* 222, 224, 254, 275.

15. *Virginia Evening Chronicle,* February 13, 1877.

16. *New York Times,* August 13, 1879; Gorham, *My Memories of the Comstock,* 190–92. It was in 1888 that Finlen fought a man over politics at the Gould & Curry dump below D Street. He hit his opponent in the back of the neck with the handle of a heavy knife. The man fell, paralyzed, and died the next day. Finlen was arrested for murder, but because testimony was inconsistent, he was acquitted. (Doten, *Journals,* vol. 3, 1703–1704, 1708.)

17. *Territorial Enterprise,* October 29, 1880.

18. H. M. Yerington, in Russell R. Elliot, with the assistance of William D. Rowley, *History of Nevada* (Lincoln: University of Nebraska Press, 1987), 165; *Gold Hill News,* November 3, 1880.

19. Alf Doten, "Some Incidents in Mackay's Life," August 3, 1902, Alf Doten Records, Series 5, Box 16, Special Collections, Getchell Library, University of Nevada, Reno.

20. *Territorial Enterprise,* January 31, 1882.

21. *New York Graphic,* November 23, 1880, in *Territorial Enterprise,* December 1, 1880.

22. *Territorial Enterprise,* January 21, 1881; Berlin, *Silver Platter,* 274, 282. The Arabian Nights allusion had been used several months earlier to describe the costliness of Mrs. Mackay's own table service and decorations at a breakfast for a select number of friends (*Nevada State Journal,* September 2, 1880).

23. *Territorial Enterprise,* January 18, 1881.

24. Berlin, *Silver Platter,* 287–89; *Territorial Enterprise,* April 6, 1881; Altrocchi, *Spectacular San Franciscans,* 200.

25. Berlin, *Silver Platter,* 289.

26. It is a quirk of history that Arthur, who appointed Mackay, had ascended to the presidency upon the assassination of James Garfield.

27. *Nevada State Journal,* February 1, 1881, May 29, 1883; Berlin, *Silver Platter,* 296–97.

28. Berlin, *Silver Platter,* 292–94; *Nevada State Journal,* June 19, 1883.

29. *Nevada State Journal,* May 3, 1883.

30. *Nevada State Journal,* June 25, July 6, 1881.

31. Lord, *Comstock Mining and Miners,* 408–10 and appendix, table II, 416; Smith, *History of the Comstock Lode,* 229–31.

32. Smith, *History of the Comstock Lode,* 271–73; Richtofen, *Comstock Lode,* 71–73; George F. Becker, in H. L. Slosson Jr., "Deep Mining on the Comstock," 4, The Bancroft Library, University of California, Berkeley; *Mining and Scientific Press,* October 9, 1880.

33. Smith, *History of the Comstock Lode,* 250–52.

34. U.S. Eleventh Census (1890), quoted in Schlesinger, *Rise of the City,* 25; Smith, "Manuscripts and notes," carton 2a, "John W. Mackay, Bonanza King," 554.

It is interesting that this seems to have been the only meeting between Smith and Mackay, yet Smith held him in such high regard that he dedicated his book on the Comstock lode to him.

35. *Territorial Enterprise,* October 11, 1895.

CHAPTER 11. *A Changing Cast*

1. Lewis, *Silver Kings,* 167–68.

2. Smith, *History of the Comstock Lode,* 211; Lewis, *Silver Kings,* 169; Alfred E. Davis, in *San Francisco Chronicle,* December 30, 1894.

3. Schlesinger, *Rise of the City,* 154–56; *San Francisco Examiner,* September 14, 1891.

4. *New York Times,* May 8, 1883; *Nevada State Journal,* May 11, 1883.

5. *San Francisco Bulletin,* n.d., quoted in Lewis, *Silver Kings,* 170–71; *St. Louis Globe-Democrat,* in *New York Times,* May 21, 1885.

6. Smith, "Manuscripts and notes," carton 2, notebook 1d, 610a.

7. George Howard Morrison, in Smith, "Manuscripts and notes," carton 2, notebook 1d, 613; Lewis, *Silver Kings,* 171–72.

8. *St. Louis Globe-Democrat,* in *New York Times,* May 21, 1885; *San Francisco Chronicle,* May 4, 1885.

9. *San Francisco Chronicle,* November 29, 1884; *New York Times,* November 27, 1884.

10. *Nevada State Journal,* February 14, 1885; *San Francisco Call,* March 17, 1885.

11. *St. Louis Globe-Democrat,* in *New York Times,* May 21, 1885; *San Francisco Chronicle,* May 3, 4, 7, 1885; Smith, "Manuscripts and notes," carton 2a, "John W. Mackay, Bonanza King," 644.

12. *New York Times,* October 16, 1885.

13. *Nevada State Jounal,* October 5, 1883, July 15, 1884; *Appletons' Cyclopedia of American Biography,* ed. James Grant Wilson and John Fiske (New York: D. Appleton & Co., 1889), vol. 4, 127, http://www.hti.umich.edu (accessed September 27, 2005); Fulton, "Reminiscences of Nevada," 85.

14. Chandler, *Visible Hand,* 12, 78, 188; Glenn Porter, *The Rise of Big Business, 1860–1920* (1973; 3rd ed., Wheeling, Ill.: Harlan Davidson, 2006), 1–4.

15. Andrew Carnegie, in Kirkland, *Dream and Thought,* 129; John Russell Young, *Men and Memories,* 444–46.

16. Chauncey M. Depew, *My Memories of Eighty Years,* chap. 22, World Wide School Library, http://www/worldwideschool.org/library/books/hst/biography/My MemoriesofEightyYears/chap25.html (accessed March 22, 2006).

17. Seitz, *James Gordon Bennetts,* 214–18, 234, 239–40, 267–70.

18. Coe, *The Telegraph,* 89.

19. Seitz, *James Gordon Bennetts,* 279–80, 341–44, 350.

20. Ibid., 340; Neville, *Fantastic City,* 118; John Steele Gordon, *A Thread Across*

the Ocean: The Heroic Story of the Transatlantic Cable (New York: Walker & Co., 2002), 210–11; New York *World,* July 22, 1902.

21. Smith, "Manuscripts and notes," carton 2a, "John W. Mackay, Bonanza King," 622–25; Maury Klein, *The Life and Legend of Jay Gould* (1986; Baltimore: Johns Hopkins University Press, 1997), 394–95.

22. Chandler, *Visible Hand,* 174, 375.

23. *New York Times,* Feb. 23, 1881; Porter, *Rise of Big Business,* 2–3.

24. Klein, *Life and Legend of Jay Gould,* 312; John Russell Young, *Men and Memories,* 444–46; Smith, "Manuscripts and notes," carton 2a, "John W. Mackay, Bonanza King," 622.

25. Porter, *Rise of Big Business,* 41; Kirkland, *Dream and Thought,* 157. See also Myers, *History of the Great American Fortunes,* 308, 482, 494; Josephson, *Robber Barons,* 192–93.

26. Chandler, *Visible Hand,* 9, 200; For an example of Gould's ownership, see Klein, *Life and Legend of Jay Gould,* 148–49. For a succinct discussion of costs and competition in early corporations, see Porter, *Rise of Big Business,* 12–16.

27. Klein, *Life and Legend of Jay Gould,* 151–52, 313–14, 364; Josephson, *Robber Barons,* 212; Myers, *History of the Great American Fortunes,* 492.

28. Klein, *Life and Legend of Jay Gould,* 153.

29. *New York Times,* August 3, 1877; Josephson, *Robber Barons,* 207.

30. *Biographical Dictionary of American Journalism,* ed. Joseph P. McKerns (New York: Greenwood Press, 1989), 579–80; Klein, *Life and Legend of Jay Gould,* 135; *New York Times,* April 1, 1888.

31. Klein, *Life and Legend of Jay Gould,* 135–36, 395–96; Morris, *Tycoons,* 142; *New York Herald,* July 21, 1902.

32. Klein, *Life and Legend of Jay Gould,* 3–8; Morris, *Tycoons,* 145.

33. *New York Times,* February 5, 1884.

34. John Russell Young, *Men and Memories,* 447; Klein, *Life and Legend of Jay Gould,* 3.

35. New York *World,* September 1, 1873, July 22, 1902; Berlin, *Silver Platter,* 299.

36. *New York Times,* February 5, 1884.

CHAPTER 12. *The New War*

1. Miller, *New History,* 275–76; John Russell Young, *Men and Memories,* 448; Chandler, *Visible Hand,* 197.

2. George P. Oslin, *The Story of Communications* (Macon, Ga.: Mercer University Press, 1992), 238; FTL Design, "A visit to the Works of Messrs. Siemens Bros.—1884," http://www.atlantic-cable.com/Article/1884Siemens/index.htm (accessed May 20, 2006). After the Titanic disaster, CS *Mackay-Bennett* was chartered to recover the bodies of those who lost their lives. Identified bodies were stored in the cable tanks, which were partly filled with ice. Bill Glover, "History of the Atlantic Cable and

Submarine Telegraphy," http://www.atlantic-cable.com/CableCos.com/CableCos/ccc.htm (accessed May 20, 2006).

3. Oslin, *Story of Communications*, 238; Pamphlet, "War of the Western Union on the Postal Telegraph," 1873[?], The Bancroft Library, University of California, Berkeley; Coe, *The Telegraph*, 91.

4. Edward L Bowen, *Legacies of the Turf: A Century of Great Thoroughbred Breeders*, vol. 1, Eclipse Press, 2003, 9, http://www.eclipsepress.com/mediaroom/pdf/legacies-vol1-ex.pdf (accessed May 22, 2006); *New York Times*, January 30, 1910.

5. Smith, "Manuscripts and notes," carton 2, "Mackay. W Nev. Bank Postal(?) Co, Cable Cos. Posted 6," section s, n.p.; *New York Times*, July 29, 1881.

6. *New York Times*, July 8, 21, August 15, 1883, May 1, 3, 1884, January 30, 1910; Klein, *Life and Legend of Jay Gould*, 313–14.

7. *New York Times*, August 15, 1883; *Nevada State Journal*, August 19, 1883.

8. *New York Times*, August 16, 29, 30, 1883; Chandler, *Visible Hand*, 199.

9. *New York Times*, August 16, 1883.

10. *San Francisco News Letter and California Advertiser*, February 19, 1887; Chandler, *Visible Hand*, 195; Smith, "Manuscripts and notes," carton 2, "Mackay. W Nev. Bank Postal(?) Co, Cable Cos. Posted 6," section s, n.p.

11. Tarbell, *Nationalizing of Business*, 41–42.

12. *New York Herald*, January 18, 1884; Klein, *Life and Legend of Jay Gould*, 196–97, 325, 331.

13. Chandler, *Visible Hand*, 10; *New York Times*, February 13, 1884.

14. Klein, *Life and Legend of Jay Gould*, 313, 325.

15. *New York Times*, July 18, 1884.

16. Smith, "Manuscripts and notes," carton 2, "Mackay. W Nev. Bank Postal(?) Co, Cable Cos. Posted 6," section t, n.p., and carton 2a, "John W. Mackay, Bonanza King," 627.

17. *New York Times*, August 27, 1884.

18. *Nevada State Journal*, August 10, 1884.

19. Klein, *Life and Legend of Jay Gould*, 334; *New York Times*, December 12, 1884.

20. *New York Times*, July 30, 1896; Klein, *Life and Legend of Jay Gould*, 379.

21. *New York Times*, October 5, 7, 26, 1884, July 30, 1885; "The Commercial Cable Company, early development as told by the president and others," http://www.cial.org.uk/cable13.htm (accessed June 10, 2007).

22. *New York Times*, July 30, 1896; Klein, *Life and Legend of Jay Gould*, 379.

23. Oslin, *Story of Telecommunications*, 238; *New York Times*, January 14, 1892, December 20, 1893.

24. Smith, *History of the Comstock Lode*, 263–64; Sam Davis, *History of Nevada*, vol. 2, 1064–66; John Russell Young, *Men and Memories*, 447–52.

25. *New York Times*, January 14, 23, 1892.

26. *New York Times,* July 22, 1875, January 1, 1885, September 14, 1887, January 23, 1892; Klein, *Life and Legend of Jay Gould,* 379–80.

27. Klein, *Life and Legend of Jay Gould,* 379–80; *New York Times,* July 11, 12, 1885.

28. *New York Times,* July 11, 12, May 25, 1885; Chandler, *Visible Hand,* 197; Klein, *Life and Legend of Jay Gould,* 379–80.

29. *New York Times,* July 11, 12, 1885.

30. *New York Times,* July 12, 1885, April 8, May 8, 1886, November 5, 1887; Klein, *Life and Legend of Jay Gould,* 383.

31. *New York Times,* January 23, 1892; Klein, *Life and Legend of Jay Gould,* 383.

32. *New York Times,* December 27, 30, 1885.

33. *New York Times,* July 30, 1896.

34. *New York Times,* October 9, 1887, July 30, 1896; Klein, *Life and Legend of Jay Gould,* 383.

35. Tarbell, *Nationalizing of Business,* 42.

36. *New York Times,* May 25, 1886; Francis L. Wellman, in Milton S. Gould, *A Cast of Hawks* (La Jolla, Calif.: Copley Books, 1985), 327.

37. *New York Times,* May 25, 1886, January 23, 1892.

38. *New York Times,* January 23, 1892; Klein, *Life and Legend of Jay Gould,* 379, 380–81, 385.

CHAPTER 13. *Near Disaster*

1. *New York Times,* October 17, 1884.

2. *New York Times,* October 18, 19, 26, December 7, 1884.

3. Smith, "Manuscripts and notes," carton 2a, "John W. Mackay, Bonanza King," 627; Lewis, *Silver Kings,* 106; John Russell Young, *Men and Memories,* 449.

4. *New York Times,* December 25, 1884.

5. *New York Times,* May 15, 1886; Smith, "Manuscripts and notes," carton 2a, "John W. Mackay, Bonanza King," 625.

6. *New York Times,* April 18, 20, 1886; *New York Herald,* May 2, 1886; Klein, *Life and Legend of Jay Gould,* 381.

7. Berlin, *Silver Platter,* 343; John Russell Young, *Men and Memories,* 449.

8. *New York Herald,* May 5, 1886; *New York Times,* May 15, 1886.

9. *New York Herald,* May 2, 5, 1886.

10. *New York Times,* May 13, 15, 1886.

11. Ibid., August 13, 16, 1886, January 27, August 7, 1887; *New York Tribune,* November 22, 1897.

12. Smith, "Manuscripts and notes," carton 2, "Mackay. W Nev. Bank Postal(?) Co, Cable Cos. Posted 6," section s, n.p.; *New York Times,* August 10, September 20, 1886; "The Commercial Cable Company, early development as told by the president

and others," http://www.cial.org.uk/cable13.htm (accessed June 10, 2007); Berlin, *Silver Platter,* 370.

13. *New York Times,* April 18–20, 1887; Smith, "Manuscripts and notes," carton 2, notebook s, n.p.

14. *New York Times,* August 5, 9–10, 1887.

15. *San Francisco Examiner,* December 10, 1894. For slightly differing accounts of how Mackay found out the Nevada Bank was in trouble, see Berlin, *Silver Platter,* 353–54; Goodwin, *As I Remember Them,* 168; John Russell Young, *Men and Memories,* 452.

16. *New York Tribune,* August 19, 1887; Berlin, *Silver Platter,* 355.

17. *New York Herald,* August 23, 1887; Smith, "Manuscripts and notes," carton 2, notebook R, n.p.; Berlin, *Silver Platter,* 355.

18. *San Francisco Chronicle,* September 14, 1887; Cross, *Financing an Empire,* vol. 1, 414.

19. *San Francisco Examiner,* December 10, 1894; Lewis, *Silver Kings,* 260; Berlin, *Silver Platter,* 354; *San Francisco Chronicle,* September 14, 1887; *New York Times,* April 26, August 4, 1887; Smith, "Manuscripts and notes," carton 2a, "John W. Mackay, Bonanza King," 645.

20. Smith, "Manuscripts and notes," carton 2a, "John W. Mackay, Bonanza King," 639–40, 645. The clerk was Jacob Levinson, who had been born in Virginia City; *San Francisco Chronicle,* September 14, 1887.

21. *San Francisco Examiner,* December 10, 1894.

22. Ibid.; Smith, "Manuscripts and notes," carton 2, notebook titled "Vol. 6 Mackay Story Duplicate," 610a; *New York Tribune,* July 21, 1902.

23. *San Francisco Chronicle,* August 28, September 14, 1887; *New York Times,* August 27, 1887.

24. *San Francisco Chronicle,* September 14, 1887; *New York Times,* September 15, 1887; John Russell Young, *Men and Memories,* 452.

25. Fair's dictation to Morrison, in Smith, "Manuscripts and notes," carton 2, notebook 1d, 612.

26. *San Francisco Examiner,* December 10, 1894; Lewis, *Silver Kings,* 262; Cross, *Financing an Empire,* vol. 1, 418.

27. *San Francisco Chronicle,* September 14, 1887; John Russell Young commented on the partnership of Mackay, Flood, and Fair: "[It was] a friendship such as we, who live outside of the atmosphere of adventure, which enfolded the Argonaut days, cannot understand" (*Men and Memories,* 443); see also Berlin, *Silver Platter,* 356.

28. *New York Times,* September 15, 1887; Cross, *Financing an Empire,* vol. 1, 418; Lewis, *Silver Kings,* 263.

29. *San Francisco Examiner,* December 10, 1894; Lewis, *Silver Kings,* 263; *New York Times,* August 28, 1887; Smith, "Manuscripts and notes," carton 2, notebook 1d, 639–40. Grant Smith said that Fair went into a wheat deal some years later and "dropped six or seven million dollars." ("Manuscripts and notes," carton 2, "Comstock Notes," section R, n.p.)

30. Smith, "Manuscripts and notes," carton 2, notebook 1d, 612.

31. *San Francisco Chronicle,* August 18, September 12, 1887.

32. *New York Times,* October 7–9, 1887, July 30, 1896; Morris, *Tycoons,* 240; Tarbell, *Nationalizing of Business,* 40.

33. *New York Times,* October 10–12, 1887, July 30, 1896; Klein, *Life and Legend of Jay Gould,* 386.

34. *New York Times,* October 9, 1887; Tarbell, *Nationalizing of Business,* 40–41.

35. *New York Times,* May 24, 1885; *San Francisco News Letter and California Advertiser,* February 19, 1887; Tarbell, *Nationalizing of Business,* 40–41.

36. *New York Times,* December 2, 1888, March 12, 1889; *San Francisco News Letter and California Advertiser,* February 19, 1887; Smith, "Manuscripts and notes," carton 2a, "John W. Mackay, Bonanza King," 634.

37. *New York Times,* February 12, 19, March 1, 1890, December 4, 1891; Tarbell, *Nationalizing of Business,* 42; Chandler, *Visible Hand,* 200. Mackay's efforts were never philosophical battles waged against big business. He established large corporations himself, while attacking entities he believed were engaged in injurious practices. Though Mackay's actions in forcing lower rates struck a blow for competition and against horizontal monopolies (only one provider of a product or service), by facilitating communication between a company's home office and distant branches, it aided corporations' efforts to create vertical monopolies, whereby one corporate entity entirely controlled various economic functions.

38. *New York Herald,* March 20, 1888; Klein, *Life and Legend of Jay Gould,* 406–9; Coe, *The Telegraph,* 89–90; Seitz, *James Gordon Bennetts,* 364.

39. *New York Tribune,* July 31, 1888; *New York Times,* July 11, 1888.

40. *New York Tribune,* July 21, 31, 1888; John Russell Young, *Men and Memories,* 450.

41. Coe, *The Telegraph,* 91; Smith, "Manuscripts and notes," carton 2a, "John W. Mackay, Bonanza King," 627; Bill Glover, "History of the Atlantic Cable and Submarine Telegraphy," http://www.atlantic-cable.com/CableCos/CCC.htm (accessed May 20, 2006); Tarbell, *Nationalizing of Business,* 42.

42. *New York Herald,* January 18, 1884.

CHAPTER 14. *Losing Control*

1. *New York Times,* February 22, 1889; Lewis, *Silver Kings,* 276.

2. *New York Times,* February 22, 1889; John Finlay, "Banks and Banking of California," *Overland Monthly and Out West Magazine* 27, no. 157 (January 1896): 87–88; King, *History of the San Francisco Stock and Exchange Board,* 212.

3. Berlin, *Silver Platter,* 366; John Russell Young, *Men and Memories,* 443.

4. Smith, "Manuscripts and notes," carton 2, "Comstock Notes," section T, n.p.; Smith, *History of the Comstock Lode,* 285; Doten, *Journals,* vol. 3, 1857.

5. Berlin, *Silver Platter,* 380; *San Francisco Examiner,* July 21, 1902.

6. Berlin, *Silver Platter*, 342; Smith, "Manuscripts and notes," carton 2, "Mackay. W Nev. Bank Postal(?) Co, Cable Cos. Posted 6," section s, n.p.

7. Smith, *History of the Comstock Lode*, 200; Smith, "Manuscripts and notes," carton 2, "Mackay. W Nev. Bank Postal(?) Co, Cable Cos. Posted 6," section s, n.p.

8. *New York Times*, July 19, 1886.

9. *San Francisco Chronicle*, December 6, 1889; Berlin, *Silver Platter*, 360, 363–66.

10. *San Francisco Chronicle*, December 6, 1889.

11. *New York Tribune*, September 28, 1891.

12. Ibid.; Lewis, *Silver Kings*, 64–65.

13. *New York Times*, January 14, 23, 1892, December 20, 1893, February 21, 1894, November 3, 1901; New York *World*, November 3, 1901; *Sacramento Daily Record Union*, December 12, 1895. The *Union* of the same date lists two other Mackay lawsuits. He lost attempting to recover $250,000 loaned to the son of an associate, and he won when a count sued for $500,000, claiming to have helped establish Commercial Cable.

14. James J. Corbett, *The Roar of the Crowd: The True Tale of the Rise and Fall of a Champion* (New York: Grosset & Dunlap, 1925), 9, 24, 27, 60; Smith, "Manuscripts and notes," carton 2, "Comstock Notes," section R, n.p.

15. Corbett, *Roar of the Crowd*, 91, 95, 200, 229, 234; Smith, "Manuscripts and notes," carton 2, "Comstock Notes," n.p.

16. Smith, "Manuscripts and notes," carton 2a, "John W. Mackay, Bonanza King," 317–18, 556.

17. Goodwin, *As I Remember Them*, 170; John Russell Young, *Men and Memories*, 441; Rollin M. Daggett, "To John W. Mackay," NC 229/12, Binder 4 (con't), n.p., Special Collections, Getchell Library, University of Nevada, Reno.

18. *New York Tribune*, February 25, 1893.

19. *New York Times*, February 25, 1893.

20. *New York Times*, February 25, May 30, 1893; *San Francisco Call*, March 22, 1893.

21. *San Francisco Call*, March 22, 1893.

22. Samuel Eliot Morison, *The Oxford History of the American People*, vol. 3 (1965; reprint, New York: Mentor Book, New American Library, 1972), 76, 82, 106; *San Francisco Call*, March 22, 1893.

23. *New York Times*, February 26, 1893; *San Francisco Call*, March 22, 1893.

24. Berlin, *Silver Platter*, 393–94; *San Francisco Call*, March 22, 1893.

25. *New York Times*, March 21, April 5, 1893; Berlin, *Silver Platter*, 394–98.

26. *New York Times*, May 30, August 11, 1893; Smith, "Manuscripts and notes," carton 2a, "John W. Mackay, Bonanza King," 317.

27. Berlin, *Silver Platter*, 273, 315, 325, 386; *Nevada State Journal*, November 15, 1883; Schlesinger, *Rise of the City*, 152.

28. *New York Times*, November 30, 1893, February 26, 1894; Berlin, *Silver Platter*, 399.

29. *New York Times,* February 26, 1894; Berlin, *Silver Platter,* 402–3; Smith, "Manuscripts and notes," carton 2, "Mackay. W Nev. Bank Postal(?) Co, Cable Cos. Posted 6," section S, n.p.

30. Little is reported regarding Mackay's having servants. Oscar Lewis said that he had a valet, urged on him by Louise in France. The young Frenchman made himself so unobtrusively useful that Mackay took him back to New York. Lewis, *Silver Kings,* 56.

31. *Mark Twain's Letters,* arranged with comment by Albert Bigelow Paine (New York: Harper & Brothers, 1917), vol. 2, 597, 614.

32. *San Francisco Examiner,* October 20, 1895; Alexander O'Grady, in Smith, "Manuscripts and notes," carton 2a, "John W. Mackay, Bonanza King," 318; *New York Tribune,* October 23, 1895.

33. Smith, "Manuscripts and notes," carton 2a, "John W. Mackay, Bonanza King," 318, 654.

34. *San Francisco Examiner,* October 20, 1895.

35. *New York Times,* October 20, 22, 1895; Berlin, *Silver Platter,* 403; *San Francisco Examiner,* October 20, 1895; Smith, "Manuscripts and notes," carton 2a, "John W. Mackay, Bonanza King," 654.

36. *San Francisco Examiner,* October 20, 1895; *New York Times,* October 21, 22, 1895.

CHAPTER 15. *Universal Eulogy*

1. *New York Tribune,* October 23, 1895; Edmond Godchaux and John Rosenfeld, in Smith, "Manuscripts and notes," carton 2a, "John W. Mackay, Bonanza King," 652–53; Berlin, *Silver Platter,* 408–9; *San Francisco Chronicle,* July 21, 1902.

2. Chandos Fulton, in *New York Times,* July 27, 1892; Edward C. Platt, in New York *World,* July 27, 1902; Smith, "Manuscripts and Notes," carton 2a, "John W. Mackay, Bonanza King," 655.

3. Berlin, *Silver Platter,* 409–10, 423–24, 427; *New York Times,* July 22, 1902; Smith, "Manuscripts and notes," carton 2, "Mackay. W Nev. Bank Postal(?) Co, Cable Cos. Posted 6," section T, n.p.

4. Smith, "Manuscripts and Notes," carton 2a, "John W. Mackay, Bonanza King," 654.

5. *New York Times,* July 22, 27, 1902.

6. Smith, "Manuscripts and notes," carton 2a, "John W. Mackay, Bonanza King," 627, and carton 2, "Comstock Notes," n.p.; Sam Davis, in *San Francisco Examiner,* July 21, 1902.

7. Andrew Carnegie, "Wealth," in Richard D. Heffner, *A Documentary History of the United States* (1952; reprint, New York: New American Library, 1965), 172–79.

8. Smith, "Manuscripts and notes," carton 2, "Comstock Notes," 9.

9. Chandos Fulton, in *New York Times,* July 27, 1892.

10. Smith, "Manuscripts and notes," carton 2a, "John W. Mackay, Bonanza King," 557.

11. P. H. Lannan, in *New York Times,* July 22, 1902; Smith, "Manuscripts and notes," carton 2, "Comstock Notes," n.p.

12. Smith, "Manuscripts and notes," carton 2a, "John W. Mackay, Bonanza King," 555.

13. R. V. Dey, quoted in *San Francisco Examiner,* July 21, 1902.

14. William Guard to Grant H. Smith, March 23, 1931, in Smith, "Manuscripts and notes," carton 2, notebook 5, n.p.

15. Smith, "Manuscripts and Notes," carton 2a, "John W. Mackay, Bonanza King," 653; "Metropolitan Opera History," http://www.metoperafamily.org/metopera/history/ (accessed June 12, 2007).

16. Smith, "Manuscripts and Notes," carton 2, "Mackay. W Nev. Bank Postal(?) Co, Cable Cos. Posted 6," section s, n.p., notebook 7, section N, n.p., carton 2a, "John W. Mackay, Bonanza King," 655; *San Francisco Call,* July 22, 1902.

17. Smith, "Manuscripts and Notes," carton 2, notebook 7, n.p.

18. *New York Herald,* September 29, November 29, 1901, January 16, March 5, May 26, 1902; Smith, "Manuscripts and notes," carton 2, notebook 7, section N, n.p.

19. Berlin, *Silver Platter,* 428.

20. Smith "Manuscripts and notes," carton 2a, "John W. Mackay, Bonanza King," 317.

21. *New York Herald,* July 21, 1902; Smith, "Manuscripts and notes," carton 2, notebook 7, section M, n.p.

22. *New York Herald,* July 21, 27, 1902; *San Francisco Examiner,* July 21, 1902; *New York Times,* July 21, 1902. See also other major American newspapers of July 21, 22, 1902.

23. Berlin, *Silver Platter,* 428–31; *New York Herald,* July 22, 1902; Coe, *The Telegraph,* 92; Smith, "Manuscripts and notes," carton 2, "Comstock Notes," 9.

24. *San Francisco Examiner,* July 21, 1902.

Bibliography

Abrams, Richard M., and Lawrence W Levine. *The Shaping of Twentieth-Century America: Interpretive Articles.* Boston: Little, Brown & Co., 1965.

Altrocchi, Julia Cooley. *The Spectacular San Franciscans.* New York: E. P. Dutton & Co., 1949.

Angel, Myron, ed. *History of Nevada.* 1881. Reprint, Berkeley, Howell-North Books, 1958.

Annual Mining Review and Stock Ledger. San Francisco: Verdenal, Harrison, Murphy & Co., 1876.

Annual Report of the Consolidated Virginia Mining Company for 1875. San Francisco: Office of the Daily Stock Exchange, 1876.

Appletons' Cyclopedia of American Biography. Vol. 4. Edited by James Grant Wilson and John Fiske. New York: D. Appleton & Co., 1889. http://www.hti.umich.edu.

Armstrong, Leroy, and J. O. Denny. *Financial California.* San Francisco: Coast Banker Publishing Co., 1916.

Atherton, Gertrude. *California: An Intimate History.* 1914. Rev. ed., New York: Horace Liveright, 1927.

———. *My San Francisco: A Wayward Biography.* Indianapolis: Bobbs-Merrill Co., 1946.

Bancroft, Hubert Howe. "Biography of William Sharon: And Material for Its Preparation, 1891." Hubert Howe Bancroft Collection, The Bancroft Library, University of California, Berkeley.

———. *Chronicle of the Builders.* San Francisco: History Co., [1890?].

247

———. "J. P. Jones." Manuscript in the handwriting of Alfred Bates. Hubert Howe Bancroft Collection, The Bancroft Library, University of California, Berkeley.

———. *The Works of Hubert H. Bancroft.* Vol. 7, *History of California, 1860–1890.* San Francisco: History Co., 1890.

———. *The Works of Hubert H. Bancroft.* Vol. 25, *History of Nevada, Colorado and Wyoming 1540–1888.* San Francisco: History Co., 1890.

Bean, Walton. *California: An Interpretive History.* New York: McGraw Hill Book Co., 1968.

Beckert, Sven. *The Monied Metropolis: New York City and the Consolidation of the American Bourgeoisie, 1850–1896.* Cambridge: Cambridge University Press, 2001.

Berlin, Ellin. *Silver Platter.* Garden City, N.Y.: Doubleday & Co., 1957.

Bethel, John D. *A General Business and Mining Directory of Storey, Lyon, Ormsby and Washoe Counties, Nevada.* Virginia City: John D. Bethel & Co., 1875.

Bierce, Ambrose. *Black Beetles in Amber.* San Francisco: Western Authors, 1892.

Biographical Dictionary of American Journalism. Edited by Joseph P. McKerns. New York: Greenwood Press, 1989.

Bowles, Samuel. *Across the Continent: A Summer's Journey to the Rocky Mountains, the Mormons, and the Pacific States with Speaker Colfax.* 1866. Reprint, Ann Arbor: University Microfilms, 1966.

Brechin, Gray. *Imperial San Francisco: Urban Power, Earthly Ruin.* Berkeley: University of California Press, 1999.

Browne, J. Ross. *A Peep at Washoe and Washoe Revisited.* 1864, 1869. Reprint, Balboa Island, Calif.: Paisano Press, 1959.

Browne, Lina Fergusson, ed. *J. Ross Browne: His Letters, Journals and Writings.* Albuquerque: University of New Mexico Press, 1969.

Bureau of the Census, Historical Statistics of the United States: Colonial Times to 1970, Part 1. Washington, D.C.: Government Printing Office, 1975.

Chandler Jr., Alfred D. *The Visible Hand: The Managerial Revolution in American Business.* Cambridge: Harvard University Press, 1977.

Chatteriee, Pratap. "The Gold Rush Legacy: Greed, Pollution and Genocide." http://www.earthisland.org/ejournal/spring98g wr.html.

Coe, Lewis. *The Telegraph: A History of Morse's Invention and Its Predecessors in the United States.* Jefferson, N.C.: McGarland & Co., 1993.

Colbert, David, ed. *Eyewitness to the American West.* New York: Penguin Putnam, 1998.

Commager, Henry Steele. *The American Mind: An Interpretation of American Thought and Character since the 1880s.* 1950. Reprint, New Haven: Yale University Press, 1966.

———. *The Era of Reform 1830–1860.* Princeton: D. Van Nostrand, 1960.

———, ed. *Documents of American History.* 1934. 7th ed. New York: Meredith Publishing Co., 1963.

Committee on Mines and Mining of the House of Representatives of the United States. *Report of The Commissioners and Evidence Taken by the Committee on*

Mines and Mining of the House of Representatives of the United States, in Regard to the Sutro Tunnel, Together with the Arguments and Report of the Committee, Recommending a Loan by the Government in Aid of the Construction of Said Work. Washington, D.C.: M'Gill & Witherow, 1872.

Corbett, James J. The Roar of the Crowd: The True Tale of the Rise and Fall of a Champion. New York: Grosset & Dunlap, 1925.

Cronise, Titus Fey. The Natural Wealth of California. San Francisco: H. H. Bancroft & Co., 1868.

Cronon, William. Changes in the Land: Indians, Colonists, and the Ecology of New England. New York: Hill and Wang, 1983.

Cronon, William, George Miles, and Jay Gitlin, eds. Under an Open Sky: Rethinking America's Western Past. New York: W. W. Norton & Co., 1992.

Cross, Ira. Financing an Empire: The History of Banking in California. 4 vols. Chicago: S. J. Clark Publishing Co., 1927.

Crowell, Max, in collaboration with Nevada State Writers' Project Work Projects Administration. A Technical Review of Early Comstock Mining Methods. Reno: Nevada State Bureau of Mines, 1941.

Current, Richard N., and John A. Garraty, eds. Words That Made American History since the Civil War. 2d ed. Boston, Toronto: Little, Brown & Co., 1962.

Dana, Julian. The Man Who Built San Francisco: A Study of Ralston's Journey with Banners. New York: Macmillan Co., 1936.

Dangberg, Grace. Carson Valley: Historical Sketches of Nevada's First Settlement. 1972. Reprint, Reno: Carson Valley Historical Society, 1979.

——. Conflict on the Carson: A Study of Water Litigation in Western Nevada. Minden, Nev.: Carson Valley Historical Society, 1975.

Davis, Sam P. The History of Nevada. 2 vols. Reno: Elms Publishing, 1913.

Davis, William Heath. Seventy-five Years in San Francisco. Orig. ed., Sixty Years in California, 1889. Reprint, San Francisco: John Howell, 1929. "Appendix I. Chinese in California," http://www.zpub.com/sf/hb75yap3.htm (accessed May 6, 2002).

Depew, Chauncey M. My Memories of Eighty Years. World Wide School Library. www.worldwideschool.org/library/books/hst/biography/MyMemoriesofEighty Years/chap.25.html (accessed March 22, 2006).

DeQuille, Dan [William Wright]. The Big Bonanza. 1876; reprint, Las Vegas: Nevada Publications, 1974.

Deverell, William, ed. A Companion to the American West. Malden, Mass.: Blackwell Publishing, 2004.

Dobie, Charles Caldwell. San Francisco: A Pageant. New York: D. Appleton-Century Co., 1943.

Doten, Alf. The Journals of Alfred Doten: 1849–1903. Edited by Walter Van Tilburg Clark. Reno: University of Nevada Press, 1973.

Drury, Wells. An Editor on the Comstock Lode. 1936. Reprint, Reno, University of Nevada Press, 1984.

Dwyer, Richard A., and Richard E. Lingenfelter. *Dan DeQuille the Washoe Giant: A Biography and Anthology*. Reno: University of Nevada Press, 1990.

Egan, Ferol. *Last Bonanza Kings: The Bourns of San Francisco*. Reno: University of Nevada Press, 1998.

———. *Sand in a Whirlwind: The Paiute Indian War of 1860*. New York: Doubleday & Co., 1972.

Eldredge, Zoeth Skinner. *The Beginnings of San Francisco: From the Expedition of Anza, 1774 to the City Charter of April 15, 1850 with Biographical and Other Notes*. New York: John C. Rankin Co., 1912.

———. *History of California*. New York: Century History Co., 1915.

Elliott, Russell R. *Servant of Power: A Political Biography of Senator William M. Stewart*. Reno: University of Nevada Press, 1983.

Elliott, Russell R., with the assistance of William D. Rowley. *History of Nevada*. Lincoln: University of Nebraska Press, 1987.

Ethington, Philip J. *The Public City: The Political Construction of Urban Life in San Francisco, 1850–1900*. 1994. Reprint, Berkeley: University of California Press, 2001.

Finlay, John. "Banks and Banking of California." *Overland Monthly and Out West Magazine* 27, no. 157 (January 1896): 87–88.

Foreman, Charles. "Statement from Charles Foreman." [1887?]. Hubert Howe Bancroft Collection, The Bancroft Library, University of California, Berkeley.

Furnas, J. C. *The Americans: A Social History of The United States 1587–1914*. New York: G. P. Putnam's Sons, 1969.

Gagey, Edmond M. *The San Francisco Stage: A History*. New York: Columbia University Press, 1950.

Galloway, John Debo. *Early Engineering Works Contributory to the Comstock*. University of Nevada Bulletin, vol. 41, no. 5. Reno: Nevada State Bureau of Mines and the Mackay School of Mines, June 1947.

Gillis, Steve. *Memories of Mark Twain and Steve Gillis*. Sonora, Calif.: Banner, n.d.

Glasscock, C. B. *The Big Bonanza*. Indianapolis: Bobbs-Merrill, 1931.

———. *Lucky Baldwin: The Story of an Unconventional Success*. N.d. Reprint, Reno: Silver Syndicate Press, 1993.

Goodwin, C. C. *As I Remember Them*. Salt Lake City: Salt Lake Commercial Club, 1913.

———. *The Story of the Comstock Lode: "Lest We Forget."* Boston: Long, Pierce & Co., n.d.

Gordon, John Steele. *A Thread Across the Ocean: The Heroic Story of the Transatlantic Cable*. New York: Walker & Co., 2002.

Gorham, Harry M. *My Memories of the Comstock*. Gold Hill, Nev.: Gold Hill Publishing, 2005.

Gould, Milton S. *A Cast of Hawks*. La Jolla, Calif.: Copley Books, 1985.

Greever, William S. *The Bonanza West: The Story of the Western Mining Rushes 1848–1900*. Norman: University of Oklahoma Press, 1963.

Hacker Louis M., and Benjamin B. Kendrick, with the collaboration of Helene S.

Zahler. *The United States since 1865.* 1932. 4th ed., New York: Appleton-Century-Crofts, 1949.

Harpending, Asbury. *The Great Diamond Hoax: And Other Stirring Incidents in the Life of Asbury Harpending.* Edited by James H. Wilkins. San Francisco: James H. Barry, 1913.

Hart, Jerome A. *In Our Second Century: From an Editor's Notebook.* San Francisco: Pioneer Press, 1931.

Heffner, Richard D. *A Documentary History of the United States.* 1952. Reprint, New York: New American Library, 1965.

Highton, Jake. *Nevada Newspaper Days: A History of Journalism in the Silver State.* Stockton, Calif.: Heritage West Books, 1990.

Hinkle, George, and Bliss Hinkle. *Sierra-Nevada Lakes.* 1949. Reprint, Reno: University of Nevada Press, 1987.

Hittell, John S. *The Commerce and Industries of the Pacific Coast.* San Francisco: A. L. Bancroft & Co., Publishers, 1882.

———. "The Mining Excitements of California." *Overland Monthly and Out West Magazine* 2, no. 5 (May 1869): 413–17.

Howard, Thomas Frederick. *Sierra Crossing: First Roads to California.* Berkeley: University of California Press, 1998.

Hulse, James W. *The Nevada Adventure: A History.* 1965. 6th ed., Reno: University of Nevada Press, 1990.

Ingram, J. S. *Centennial Exposition Described and Illustrated.* Philadelphia: Hubbard Bros., 1876. http://fax.libs.uga. edu.

Jackson, Kenneth T., and David S. Dunbar, eds. *Empire City: New York Through the Centuries.* New York: Columbia University Press, 2002.

James, George Wharton. *The Lake of the Sky: Lake Tahoe in the High Sierras of California and Nevada.* 1915. 2d ed., Pasadena, Calif.: Radiant Life Press, 1921.

James, Ronald M. *The Roar and the Silence: A History of Virginia City and the Comstock Lode.* Reno, University of Nevada Press, 1998.

James, Ronald M., and C. Elizabeth Raymond, eds. *Comstock Women: The Making of a Mining Community.* Reno: University of Nevada Press, 1998.

Josephson, Matthew. *The Robber Barons: The Great American Capitalists 1861–1901.* 1934. Reprint, New York: Harcourt, Brace & World, A Harvest Book, 1962.

Kelly, J. Wells. *First Directory of Nevada Territory.* San Francisco: Valentine Co., 1862.

King, Joseph L. *History of the San Francisco Stock and Exchange Board, by the Chairman, Jos. L. King.* San Francisco: J. L. King, 1910.

Kirkland, Edward Chase. *Dream and Thought in the Business Community, 1860–1900.* 1956. Reprint, Chicago: Ivan R. Dee, 1990.

Klein, Maury. *The Life and Legend of Jay Gould.* 1986. Reprint, Baltimore: Johns Hopkins University Press, 1997.

Kowalewski, Michael, ed. *Gold Rush: A Literary Exploration.* Berkeley: Heyday Books, 1997.

Lavender, David. *Nothing Seemed Impossible: William C. Ralston and Early San Francisco.* 1975. Reprint, Palo Alto, Calif.: American West, 1981.

Leopold, Richard W., and Arthur S Link. *Problems in American History.* Englewood Cliffs, N.J.: Prentice-Hall, 1957.

Leslie, Mrs. Frank. *California: A Pleasure Trip from Gotham to the Golden Gate, April, May, June, 1877.* Reprint, Nieuwkoop, Netherlands: B. De Graaf, 1972. http://members.door.net/nbclumber/Leslie/Ch32.htm (accessed June 11, 2002).

Lewis, Oscar. *The Big Four.* 1938. Reprint, New York: Ballantine Books, Comstock Ed., 1971.

———. *High Sierra Country.* 1955. Reprint, Reno: University of Nevada Press, 1984.

———. *Silver Kings: The Lives and Times of Mackay, Fair, Flood and O'Brien, Lords of the Nevada Comstock Lode.* New York: Alfred A Knopf, 1947.

———, ed. *This Was San Francisco: Being First-Hand Accounts of the Evolution of One of America's Favorite Cities.* New York: David McKay Co., 1962.

Limerick, Patricia Nelson. *The Legacy of Conquest: The Unbroken Past of the American West.* New York: W. W. Norton & Co., 1987.

Linderman, H. R. *Report of the Director of the Mint to the Secretary of the Treasury for the Fiscal Year Ended June 30, 1875.* Washington, D.C., Government Printing Office, 1875.

Lingenfelter, Richard E. *The Hardrock Miners: A History of the Mining Labor Movement in the American West 1863–1893.* Berkeley: University of California Press, 1974.

———, with an introduction by David F. Myrick. *The Newspapers of Nevada: A History and Bibliography, 1858–1958.* San Francisco: John Howell—Books, 1964.

Lloyd, B. E. *Lights and Shades in San Francisco.* 1876. Facsimile ed., Berkeley: Berkeley Hills Books, 1999.

Lord, Eliot. *Comstock Mining and Miners.* 1883. Reprint, Berkeley: Howell-North, 1959.

Lyman, George. *Ralston's Ring.* 1937. New York: Ballantine Books, Comstock Ed., 1971.

———. *The Saga of the Comstock Lode.* New York: Charles Scribner's Sons, 1934. Reprint, New York: Ballantine Books, Comstock Ed., 1971.

Lyttleton, George. *As I Recall Them: Memories of Crowded Years.* New York: Wilson-Erickson, 1936.

Mack, Effie Mona. *Nevada: A History of the State from the Earliest Times through the Civil War.* Glendale, Calif.: Arthur H. Clark Co., 1936.

Marye Jr, George Thomas. *From '49 to '83 in California and Nevada: Chapters from the Life of George Thomas Marye, a Pioneer of '49.* San Francisco: A. M. Robertson, 1923.

Mathews, Mrs. M. M. *Ten Years in Nevada, or Life on the Pacific Coast.* 1880. Reprint, Lincoln: University of Nebraska Press, 1985.

McClellan, R. Guy. *The Golden State: A History of the Region West of the Rocky Mountains; Embracing California, Oregon, Nevada, Utah, Arizona, Idaho, Wash-*

ington Territory, British Columbia, and Alaska, from the Earliest Period to the Present Time . . . Philadelphia: William Flint & Co., 1872.

McDonald, Douglas. The Legend of Julia Bulette and the Red Light Ladies of Nevada. Edited by Stanley Paher. Las Vegas: Nevada Publications, 1980.

Michelson, Miriam. The Wonderlode of Silver and Gold. Boston: Stratford, 1934.

Miller, William. A New History of the United States. New York: George Braziller, 1958.

Morison, Samuel Eliot. The Oxford History of the American People. Vol. 3. 1965. Reprint, New York: Mentor Book, New American Library, 1972.

Morison, Samuel Eliot, and Henry Steele Commager. The Growth of the American Republic. 2 vols. 1930. Rev. ed., New York: Oxford University Press, 1958.

Morris, Charles R. The Tycoons: How Andrew Carnegie, John D. Rockefeller, Jay Gould, and J. P. Morgan Invented the American Supereconomy. New York: Henry Holt & Co., 2005.

Myers, Gustavus. History of the Great American Fortunes. 1907. Reprint, New York: Random House, 1964.

Neville, Amelia Ransome. The Fantastic City: Memoirs of the Social and Romantic Life of Old San Francisco. Edited and revised by Virginia Brastow. Boston and New York: Houghton Mifflin Co., 1932.

O'Connor, Richard. Gould's Millions. New York: Doubleday & Co., 1962.

Oslin, George P. The Story of Communications. Macon, Ga.: Mercer University Press, 1992.

Ostrander, Gilman M. Nevada: The Great Rotten Borough, 1859-1964. New York: Knopf, 1966.

Parrington, Vernon L. Main Currents in American Thought. Vol. 2, The Romantic Revolution in America, 1800-1860. New York: Harvest Book, Harcourt, Brace & World, 1927.

Paul, Rodman. The Far West and the Great Plains in Transition, 1859-1900. New York: Harper & Row, 1988.

———. Mining Frontiers of the Far West, 1848-1880. New York: Holt, Rinehart, & Winston, 1963.

Phillips, Kevin. Wealth and Democracy: A Political History of the American Rich. New York: Broadway Books, 2002.

Pomeroy, Earl. The Pacific Slope: A History of California, Oregon, Washington, Idaho, Utah and Nevada. New York: Alfred A. Knopf, 1966.

Porter, Glenn. The Rise of Big Business, 1860-1920. 1973. 3rd ed., Wheeling, Ill.: Harlan Davidson, 2006.

Powell, John J. Nevada: The Land of Silver. San Francisco: Bacon & Co., 1876.

Raymond, Rossiter W. Silver and Gold: An Account of the Mining and Metallurgical Industry of the United States. 1873. Reprint, University of Michigan: Scholarly Publishing Office, n.d.

———. Statistics of Mines and Mining in the States and Territories West of the Rocky Mountains. Washington, D.C.: Government Printing Office, 1877.

Richtofen, Ferdinand Bacon. *The Comstock Lode: Its Character, and the Probable Mode of Its Continuance in Depth.* San Francisco: Towne & Bacon, 1866.

Robbins, William G. *Colony and Empire: The Capitalist Transformation of the American West.* Lawrence: University Press of Kansas, 1994.

Robinson, Forrest G., ed. *The New Western History: The Territory Ahead.* Tucson: University of Arizona Press, 1997.

Rocha, Guy Louis. "The Many Images of the Comstock Miners' Union." http://www.nevadalabor.com/rocha.html (accessed May 17, 2002).

Rowley, William D. *Reclaiming the Arid West: The Career of Francis G. Newlands.* Bloomington: Indiana University Press, 1996.

Roy, William G. *Socializing Capital: The Rise of the Large Industrial Corporation in America.* Princeton, N.J.: Princeton University Press, 1997.

Schlesinger, Arthur Meier. *The Rise of the City, 1878–1898.* 1933. Reprint, New York: Macmillan Co., 1969.

Scott, E. B. *The Saga of Lake Tahoe.* Crystal Bay, Lake Tahoe, Calif.: Sierra-Tahoe Publishing Co., 1957.

Seitz, Don C., *The James Gordon Bennetts, Father and Son: Proprietors of the New York Herald.* Indianapolis: Bobbs-Merrill Co., 1928.

Shinn, Charles Howard. "California Mining Camps." *Overland Monthly and Out West Magazine* 4, no. 20 (August 1884): 173–75.

———. *The Story of the Mine: As Illustrated by the Great Comstock Lode of Nevada.* 1910. Reprint, Reno: University of Nevada Press, 1980.

Smith, Grant H. *The History of the Comstock Lode: 1850–1920.* 1943. Reprint, Reno: University of Nevada Press, 1998.

———. "Manuscripts and notes chiefly concerning the Comstock Lode 1859–1936." Carton 2. The Bancroft Library, University of California, Berkeley.

———. "Manuscripts and notes chiefly concerning the Comstock Lode 1859–1936." Carton 2a. "John W. Mackay, Bonanza King." The Bancroft Library, University of California, Berkeley.

Stewart Jr., Robert E., and Mary Frances Stewart. *Adolph Sutro: A Biography.* Berkeley: Howell-North, 1962.

Stewart, William M. *Reminiscences of Senator William M. Stewart of Nevada.* Edited by George Rothwell Brown. New York: Neale Publishing Co., 1908.

———. Stewart Manuscript Collection. Nevada Historical Society, Reno, Nevada.

Stone, Irving. *Men to Match My Mountains: The Opening of the Far West, 1840–1900.* New York: Doubleday & Co., 1956.

Sutro, Adolph. "Autobiographical Notes." 1890. The Bancroft Library, University of California, Berkeley.

———. "Closing Argument of Adolph Sutro." April 22, 1872, 15. The Bancroft Library, University of California, Berkeley.

Takaki, Ronald. *A Different Mirror: A History of Multicultural America.* Boston: Little, Brown & Co., 1993.

Tarbell, Ida M. *The Nationalizing of Business, 1878–1898.* A History of American Life, vol. 9. 1927. Reprint, New York: Macmillan Co., 1969.

Tilton, Cecil G. *William Chapman Ralston, Courageous Builder.* Boston: Christopher Publishing House, 1935.

Tinkham, George H. *California: Men and Events.* Stockton, Calif.: Record Publishing Co., 1915.

Townley, John M. "Reclamation in Nevada 1850–1904." PhD diss., University of Nevada, 1976.

Twain, Mark. *Mark Twain's Letters.* Vol. 2. Arranged with comment by Albert Bigelow Paine. New York: Harper & Brothers, 1917.

———. *Roughing It.* 2 vols. 1871. Reprint, New York: Harper & Brothers, 1913.

Upshur, George Lyttleton. *As I Recall Them: Memories of Crowded Years.* New York: Wilson-Erickson, 1936.

Waldorf, John Taylor. *A Kid on the Comstock: Reminiscences of a Virginia City Childhood.* Edited, with introduction and commentary, by Dolores Waldorf Bryant. 1968. Reprint, Reno: Vintage West Series, University of Nevada Press, 1991.

Washow, Robert Irving. *Jay Gould: The Story of a Fortune.* New York: Greenberg, Publisher, 1928.

Wells, Harry L. "Gold Lake: The First Stampede in the California Mines." *Overland Monthly and Out West Magazine* 4, no. 23 (November 1884): 519–25.

West, Richard Samuel. *The San Francisco Wasp: An Illustrated History.* Easthampton, Mass.: Periodyssey Press, 2004.

Whitehill, Henry R. *Biennial Report of the State Mineralogist of the State of Nevada for the Year 1871 and 1872.* Carson City: John J. Hill, state printer, 1872.

———. *Biennial Report of the State Mineralogist of the State of Nevada for the Year 1873 and 1874.* Carson City: John J. Hill, state printer, 1874.

———. *Biennial Report of the State Mineralogist of the State of Nevada for the Year 1875 and 1876.* Carson City: John J. Hill, state printer, 1876.

Williams, Henry T. *The Pacific Tourist.* New York: Henry T. Williams, 1877.

Wittemore, Robert Clifton. *Makers of the American Mind.* New York: William Morrow & Co., 1964.

W.P.A. Writer's Project. *Individual Histories of the Mines of the Comstock.* A joint project of the W.P.A. Writer's Project and the Nevada State Bureau of Mines. Reno: Nevada State Bureau of Mines, 1942[?].

Yellow Jacket Silver Mining Company Records 1861–1911. Vol. 1. Special Collections Dept., Getchell Library, University of Nevada, Reno.

Young, John P. *Journalism in California.* San Francisco: Chronicle Publishing Co., 1915.

Young, John Russell. *Men and Memories.* Edited by May D. Russell Young. New York: F. Tennyson Neely, 1901.

Zinn, Howard. *A People's History of the United States.* 1980. Reprint, New York: HarperPerennial, 1990.

Index

Adams, Edwin, 124
Alexander III, 120, 143
Alta Mine, 135, 140
American Asiatic Association, 209
American Rapid Telegraph, 164, 168–69, 171, 172–73
Ames, Oakes, 83–94
Anderson, James, 56
Angel, Myron, 80, 228n25
Anglo-American Company, 176, 178
Arthur, Chester, 120, 125, 143, 237n26
Ashburner, William, 64
Associated Press, 153–54, 156, 160
Atherton, Gertrude, 127
Auction Lunch Saloon, 30, 52, 90
Aurora, 24, 28
Austin, W. Wallace, 86–87

Bacon Mill, 44, 71
Balch, William Ralston, 121
Bald Peak, 11
Baldwin, E. T. "Lucky," 78, 85
Baltimore and Ohio Railroad, 164
Baltimore and Ohio Telegraph Company:
    American Rapid Telegraph and, 164;
    Baltimore and Ohio Railroad and, 164;

Bankers and Merchants' Telegraph
    Company and, 158, 170; as competitive
    enterprise, 158, 164–66, 186–87; Garrett
    family and, 166; Mackay and, 164–65, 166,
    168, 172; J. P. Morgan and, 186; New York
    and, 164, 171–72; and rate reductions, 168;
    syndicate of, 172; telegraph operators' union
    and, 162; Western Union and, 171. *See also*
    Garrett, Robert
Bancroft, H. H., 40–41, 148
Bankers and Merchants' Telegraph Company:
    American Rapid Telegraph and, 164; as
    competitive enterprise, 158, 164–65; Jay
    Gould and, 169–70; Mackay and, 164–66,
    167–73; and stock panic, 164; Edward
    Stokes and, 167–73; Western Union and,
    169–73. *See also* Dimrock, A. W.
Bank of California: Bonanza Firm and, 82,
    87, 90, 92; Comstock and, 39, 49–50, 52,
    93; description of, 39, 75, 87; failure of, 79,
    89–93, 105, 183; and fire of 1875, 95; Mackay
    and attorneys for, 49; Louis McLane
    and, 82; mills and, 52; D. O. Mills and,
    89; Nevada Bank of San Francisco and,
    182–83; William Ralston and, 75, 79, 87, 90;
    William Sharon and, 6, 39, 52, 81, 91

Bank Ring (the Ring): associates of, 25, 27, 32, 33, 39; Bonanza Firm and, 43–44, 79, 82; description of, 2; James Fair and, ix; *Gold Hill News* and support of, 45; Hale & Norcross mine and, 6–7, 34, 41, 43; Mackay and, ix, 2, 81–92; management of, 112; mills of, 7, 39, 40, 52; monopoly and, 40, 49, 53, 60, 81–82; William Sharon and, 32, 34, 40, 49, 52, 79–81; Adolph Sutro and, 52; and water company, 61; and water rights, 59–60; weakness of, 44; and Yellow Jacket mine, 39–40

Barbary Coast (San Francisco), 23, 74

Barnes, Zink, 23, 31

Barron, Edward, 33, 65

Bay and Coast Telegraph, 178

Becker, George F., 144

Belcher mine, 61, 65, 80, 139

Bell, Thomas, 76, 90

Bennett, James Gordon Jr.: biographical information on, 151–53; cable built by, 174–79, 189; cable repair ship named for, 160, 166; Jay Gould and, 150, 156, 157, 163, 188–89; as host, 142–43; and Mackay against Jay Gould, 150, 174–79, 181, 189; Mackay as ally of, 154, 157; Mackay as partner of, 162, 166, 174–79, 205; on meeting Mackay, 143, 151, 153; and *New York Herald* editorship, 143, 151, 152–53; and using *New York Herald* to attack Jay Gould, 163, 181, 188–89; and using *New York Herald* to support Mackay, 177, 181; and percentage of cable company, 157; and plans for cable company, 162; and second cable, 166

Bennett, James Gordon Sr., 9, 48–49, 143, 151, 152

Berlin, Ellin, 120, 143

Best & Belcher mine, 65, 68, 99

Bierce, Ambrose, 194

Big Bonanza: and access through Gould & Curry mine, 50; development of, 2; discovery of, 2; estimations of value of, 85; James Fair and, 54, 55; Mackay and, 78, 88, 97; Mackay referred to as, 149; Helena Modjeska and, 55; original owner of, 225n18; and shelter built above it, 18; Sierra Nevada mine and, 135–36, 137; Johnny Skae and, 136; and stock market, 118

Bland-Allison Bill, 84

Board of Trade and Transportation, 177

Bonanza Firm: attacks and, 100, 112–16, 180, 193; attorney for, 130; Bank of California challenged by, 82, 87, 90, 92; Bank Ring and, 82; California mine organized by, 75; and C & C shaft, 77; and Centennial Exposition, 105–6, 111; Con. Virginia mine and management by, 31, 63, 65, 67; description of members of, 104, 105, 180; S. P. Dewey and, 113–16, 129; James Fair and, 54, 56, 102, 105, 135, 148; and fire of 1875, 97; James Flood and, 45, 92, 106–7, 127, 130; and flumes, 101; Hale & Norcross mine management by, 44; and litigation, 128–30; John Mackay and, 45, 54, 102, 105; and milling operations, 81; mine heat and, 70; mine management by, 67–68, 112, 113, 129; mine production managed by, 54; Nevada Bank of San Francisco and, 87, 93; newspaper defense of, 117; William O'Brien and, 127; and the press, 117–18; William Ralston and, 87, 92; Hermann Schussler and, 71; Sierra Nevada mine and, 138–39; stamp mills and, 77, 129; stock management by, 107–8, 111–12; stock ownership by, 71; Edward Stokes and, 125; superintendent of mills of, 81; and taxes, 80–81, 128–29; timber operations of, 81, 101, 129; Union mine and, 137, 139; and would-be assassin, 198; yield of, 80

Bonanza King(s): Bank of California and, 90; and commissions, 89; description of, 88; James Fair as, 55; Mackay as, 141; Ralston and, 230n21

Bonner, John, 197–98

Bonynge, C. W., 106, 192–94

Borglum, Gutzon, 211

Bowers, Eilley Orum, 107, 232n13

Bradley, E. C., 126

Brander, George L., 149, 181–82, 184–85

Brown, Thomas, 182–83

Browne, J. Ross, 19, 22, 59, 105, 217n11

Bryant, Dr. Edmond, 15, 35–36, 37

Bryant, Marie Louise Antoinette Hungerford. *See* Mackay, Louise

Bull, Alpheus, 52

Bullion mine, 27–28, 33, 52–53, 71

Burke, Jack, 195

Burke, John H., 130

Burleigh drills, 77

California Constitution, 139

California mine: and C & C shaft, 77, 99; description of, 18, 63, 75; James Fair and, 99; and fire of 1875, 95, 96; Mackay and, 87; national Centennial Exposition and, 105–6;

production of, 75, 113, 144; Russell Sage on, 163; stocks of, 85, 113

California Mining Company, 75

California Theater, 75, 124

Canadian Pacific Railroad, 116, 178, 205

Canadian Pacific Telegraph, 178

C & C (mine shaft), 77, 99

Carnegie, Andrew, 206–7

Carson City, 19, 23, 62, 96, 98, 160

Carson River, 24, 40, 51, 59, 115

Centennial Exposition, 105, 233n27

Central mine, 68, 75

Central No. 2 mine, 68, 75

Central Valley (Calif.), 61, 75

Chandler, Albert Brown, 167, 187

Chollar-Potasi (mine): Bullion mine and, 27; development in, 61; James Fair and, 96; and fire of 1875, 96; Hale & Norcross mine and, 41; and joint shaft, 107; litigation and, 25; John Mackay and, 26, Milton mine and, 26

Clews, Henry, 162

Cole, Dr. A. M., 103–4

Cole Mining Company, 60

Colonna, Don Ferdinand, 149, 201

Combination Shaft, 232n14

Commercial Cable: and cable construction, 165, 174–75, 189; employees and stock, 206; Jay Gould and, 181; incorporated as, 164; John Mackay Jr. as board member of, 200, 202–3; Pacific cable and, 209–10; Postal Telegraph and, 205; and rates, 176, 177, 179, 186, 189; Henry Rosener and, 182; service by, 175; Edward Stokes and, 187; George Ward and, 164, 175, 176

Commercial Cable Building, 205

Comstock: Bank of California and, 50, 91; Bank Ring and, 53; Zink Barnes on, 23; J. Ross Browne on, 22, 59; business owners and, 51; characters of, 49, 50, 55; and charges regarding wheat deal, 180; Con. Virginia mine and, 64, 69, 71, 75; Sam Curtis and, 66; Rollin Daggett and, 196; decline of, 142, 144; description of, 22, 59, 74; development of, 24; Alf Doten and, 141; employment in mines of, 192; James Fair and, 31, 102, 106, 135, 137, 140, 147, 148; James Fair/Mackay partnership and, 53, 185; James Fair mining on, 31, 167; James Fair viewed by citizens of, 103, 112, 134; Theresa Fair and, 108; the Firm and, 6, 61, 68, 71, 113, 129; James Flood and, 192; James Flood and William O'Brien and, 30; gam-

bling on, 26, 134; Fred Hart and, 140–41; Hispanics in, 20; Indians and, 20; J. P. Jones and, 33, 145; Lake Tahoe and, 61; Eliot Lord and, 27; Mackay abandonment of, 145; Mackay and, 4, 68, 70, 86, 99, 102, 117, 123–24, 134–35; 150, 190; Mackay and boxing on, 194; Mackay and boys of, 103; and Mackay charity, 207; Mackay and control of, 39, 78–79; Mackay house on, 5; Mackay and mining on, 24, 31, 53, 87, 167; Mackay as outstanding character of, 2, 103; Mackay and saga of, 30; Mackay and supporters on, 6, 199; Mackay as timber worker on, 20; and Mackay riches taken from, 205; Mackay spending on, 168; Louise Mackay and, 36, 47, 73, 108, 193; Father Manogue and, 36; M. M. Mathews and, 38–39; George Mayre Jr. and, 33; Arthur McEwen on, 26; mills and, 39, 129; mine litigation on, 130; miners and, 31, 41, 56, 76, 113; miners' abandonment of, 145; mines of, 46, 50, 52–53; mining on, 20–21, 26, 41, 50; and mining wealth, 54; mining yield on, 144; Helena Modjeska and, 55; monopoly and, x, 2, 53; north end of, 63–64; opium use on, 36; ore of, 18, 64, 68, 123, 135; people arriving at, 22; problems of, 93; Pyramid Lake and, 19; William Ralston's death and, 91; George Roberts and, 112, 157; Henry Rosener and, 157; Russell Sage and, 162–63; William Sharon and, 39, 52, 68, 80, 117; Sierra Nevada mine and people of, 138–39; Johnny Skae and, 137; Grant Smith and, 3; stock market and, 28, 64, 66, 89, 136; Adolph Sutro and, 50; teamster of, 203; *Territorial Enterprise* and, 132; Mark Twain and, 202; Virginia & Truckee Railroad and, 52; volatility of, 26; J. M. Walker and, 33; water and, 59–60, 70–71; wood industry and, 81; worker on, 57

"Comstocker(s)": C. W. Boynge as, 192; as citizens of the Comstock, 51, 135, 136; James Fair and, 135; Will Gillis as, 56; Mackay and, 4, 135, 207; way of life of, 26

Comstock Lode: Bank Ring and, 2; Bonanza Firm and, 113; books on, 3; Con. Virginia and, 63–64, 67–69; Sam Curtis and, 66–68; Phillipp Deidesheimer and, 21; description of, 25, 52; James Fair and, 47, 53, 54, 67–69; James Flood, William O'Brien, and, 54; Hale & Norcross mine and, 47; location of, 16; Mackay and mining of, ix, 53, 67–70, 144; Mackay and word of, 15; Mexican mine and, 20; miners and, 41; as mining laboratory, 58; 1881 production of, 144; scientists on, 64, 144; William Sharon and, 39, 40; Sierra Nevada mine and, 135–36; stock market and, 66; Sutro tunnel and, 50; Union tunnel and, 20

Consolidated Virginia (Con. Virginia) mine: annual report and, 99, 112; Bonanza Firm and, 65, 76, 113; and Centennial Exposition, 106; and clearing titles, 130; as compared to Sierra Nevada mine, 137, 139; Sam Curtis and, 66–68; deep mine exploration and, 145; Dan DeQuille on, 63, 69, 72, 78, 85; description of, 63–64, 67, 72, 75, 99; dividends of, 144; estimations of value of, 85, 88–89; James Fair on, 99–100, 105, 112; James Fair as superintendent of, 31, 69, 76–77, 104; and fire of 1875, 94–99; James Flood buying shares of, 64, 66, 72, 85, 107; James Flood versus S. P. Dewey and, 113–17, 130; James Keene and, 107, 112, 193; location of, 64; Mackay abandonment of, 146; Mackay and description of riches, 70, 78; Mackay and development of, 71; Mackay investment and, 68, 71; Mackay management of, 69, 76–77, 85; Mackay ownership of, 4; Mackay vs. S. P. Dewey and, 114–17; Mackay and richer veins of, 107; Mackay stocks and, 87, 89, 104; and mills, 81; Nevada Block and, 137; newspapers on, 66, 77, 107, 112, 117; ore samples and, 136; Russell Sage on, 163; shaft building of, 142; William Sharon and, 68, 90–91; sledding and, 103; stock excitement and, 84–86; stock increases and, 71–72; stock losses in 1875 of, 90–91, 97; stock losses in 1876 of, 108, 113; stock prices in 1871 of, 64; stock prices in 1872 of, 66; stock prices in 1877 and, 117–18; stock success in 1873 of, 71, 78; stock success in 1875 of, 77, 105; Tiffany's of New York and, 123; work in, 65, 67–68, 69, 71, 75–77, 85; yield of, 75, 77, 100, 129

Cook, Dr. Frederick, 152

Corbett, Gentleman Jim, 195–96

Cosser, John, 53–54

Crédit Mobilier, 83–84

Cross, Ira B., 184

Crown Point mine, 29, 30, 61, 64–66

CS *Mackay-Bennett* (ship), 160, 166, 239n2

Curtis, Capt. Samuel T.: and Con. Virginia mine, 65–69; fire of 1875 and, 95–96, 98;

miners war and, 135; scandals and, 226n35; William Sharon and, 84

Daggett Rollin: on the Comstock, 23; description of, 12, 141; on Mackay, 12, 141–42, 196–97; and old western history, 3; and Sharon, 76; and *Territorial Enterprise,* 76

Davis, Sam, 1–4, 41, 53, 81

Deane, Coll, 85–86

DeGroot, Henry, 64–65

Deidesheimer, Philipp, 21, 85

DeLong, Charles, 86, 91

DeQuille, Dan: on Comstock water, 60, 70; on Con. Virginia mine, 63, 69, 72, 78, 85; on James Fair, 45; on Mackay, 54; Mackay's comments to, 87; on Mackay and Sam Curtis, 69; on Mackay and stock investments, 86; on mining, 66; and old western history, 3; Mackay pension and, 4; on Sierra Nevada mine, 136; on stock investments, 85

Dewey, Squire P.: assault on Bonanza Firm and, 111; description of, 113; claims of, 113–16, 129–30; James Flood and, 113–14; Mackay and, 114–16; and newspapers' reactions, 116–17, 130; San Francisco *Chronicle* and, 112–13; stockholders and, 115–16, 129–30; suit by, 129–30

Dey, R. V.: description of, 109; Mackay and, 199, 200, 210; on Mackay's wealth, 206, 208; Louise Mackay and, 109, 200

Dimrock, A. W., 164, 165–66

Direct United States Company, 164, 176

Divide, The, 22, 24, 27, 64, 95,

Donohue, Charles, 168, 169–71, 173

Doten, Alf: on Eilley Orum Bowers, 232n13; on cholera on Comstock, 93; on Con. Virginia mine, 67–68; on election of 1880, 140; on fire of 1875, 95; and *Gold Hill News,* 45; on Fred Hart and Mackay, 140–41; and other newspapers, 96; William Sharon and, 222n31; on Sierra Nevada mine stock panic, 138; on Johnny Skae and party, 136; on Sutro tunnel, 51

double jacking, 67, 226n25

Downie, William, 13

Downieville, 11–14, 20, 35–36, 134

Dresbach, William, 181, 183–86

Drury, Wells, 26

Duer, Katherine (Mrs. Clarence Mackay), 111, 205

Durgan, Flats, 11, 13

Edgecomb, Zeke, 60
Edmunds, George F., 164
Enterprise Publishing Company, 76
Erhardt, Joel B., 161–62
Esmeralda Claim, 24
Eureka, 11–12
Eyre, Col. Edward, 89

Fair, James G. ("Slippery Jimmy"): and alcohol, 106, 135; and Bank of California failure,
91; boastfulness of, 31, 32, 34, 44, 54, 104–5;
and Big Bonanza, 50, 54, 55; and Bonanza
Firm, 33–34, 43, 56, 129; and Burleigh drills,
77; and California mine, 99; and children,
103; and Comstock Lode, 47, 53, 54, 67–69;
Comstock vein and study by, 65; Con.
Virginia mine annual report by, 99–100, 112;
Con. Virginia mine building by, 67; as Con.
Virginia mine information source, 99–100,
105, 112, 114; and Con. Virginia mine
purchase, 64–65; as Con. Virginia mine
superintendent, 31, 69, 76–77, 104, 112, 116;
Dan DeQuille and, 45, 72; descriptions of,
31–32, 53–54, 103–5; and divorce, 147–48;
and employees, 55–57; and fire of 1875, 94,
96–99; on James C. Flood, 148, 150, 191,
236n4; James C. Flood and feelings of, 135,
148; James C. Flood and William O'Brien
and assistance by, 33; and Gold Hill News,
45; and Hale & Norcross mine, 42–45,
47, 62; Fred Hart on, 140–41; infidelity of,
147; and illness, 135; investments of, 147,
148, 180; as a larger-than-life character, 49;
and leadership abilities, 135; machinery
and, 219n47; Mackay on, 137; and Mackay
achievements, 57, 58; Mackay and assaults
by, 111–12, 148, 180–81; Mackay and assistance by, 79; and Mackay comparisons,
102, 180; and Mackay competition, 53; and
Mackay friendship, 134, 140–41; and mills,
115; mine infrastructure building by, 76;
mine litigation and, 130; as mine supervisor,
3, 31, 66, 135, 144, 167; and Nevada Bank of
San Francisco, 82, 144, 148–50, 168, 183–85;
pride of, 55; and resignation, 135, 144; and
riding a flume, 101–2; Theresa Rooney
Fair and, 32, 147–48; and partnership, 3;
William Sharon and, 50; and Sierra Nevada
mine, 61, 137; Grant Smith on, 31, 34, 54;
and social affairs, 49; stock market war and,
108; as unfair, 103; and U.S. Senate, 140;
as U.S. senator, 76, 146; and vacation, 68,

106; and water company, 61; and wealth and
estimates concerning, 78
Fair, Theresa: description of, 32, 37, 108; and
divorce, 147–48; James Fair and, 32, 147;
James Fair and accusations by, 147; Mackay
and, 148; as matchmaker, 37; on Louise
Mackay, 109; newspapers on, 147–48
Fair, Virginia, 111
Faraday (ship), 174–75
Farnsworth, Gen. J. G., 168–69, 172–73
Far West, 3, 75
Finlen, Miles, 139, 237n16
Finney (Fenimore), James ("Old Virginia"/
"Old Virginny"), 17
Firm, the. See Bonanza Firm
Fire Company No. 4, 94
Fish, Charles H., 113
Fisk, F. E., 41
Fisk, James ("Jubilee" Jim), 83, 124–25, 166–67,
171
Fleishacker, Mortimer, 207
Flood, James Clair: and Auction Lunch
Saloon, 30; Bank of California failure
and, 90, 92–93; Bank Ring and, 33, 79;
and Bonanza Firm, 3, 33, 43, 45, 88, 129;
brokerage office of, 33; J. Ross Browne
on, 105; Centennial exposition ore and,
106; Comstock vein and study by, 65;
Con. Virginia mine building by, 67; Con.
Virginia mine purchase and, 64–65; and
Con. Virginia mine stocks, 65, 66–67, 72,
85, 130, 137; death of, 191, 192; descriptions
of, 45, 88, 105, 127, 191–92; Squire P. Dewey
and, 113–15, 130; and employees, 62; James
Fair on, 54, 105, 135, 148, 150, 184, 191; James
Fair assistance by, 33; James Fair and scorn
of, 146, 148; generosity of, 126–27, 191; and
Hale & Norcross mine, 43–44; house of,
108; and illness, 150, 181, 183–84, 191–92;
James Keene and, 33, 107, 160; Mackay and,
30, 34, 52, 89, 114, 137, 144, 148, 167, 168,
192; and mills, 97, 129; and mine workings, 135, 236n4; and Nevada Bank of San
Francisco, 82, 127, 144, 149–50, 181–86;
William O'Brien and, 30, 34, 127, 128;
William Ralston and, 75, 87, 90, 92–93; and
riding a flume, 101–2; William Sharon and,
97; and Sierra Nevada mine, 137; Johnny
Skae and, 136; stock attacks and reaction by,
106–8, 113; stock exchange and influence of,
105; stock jobbing accusations and, 72; and
Union mine, 137; J. M. Walker and, 30, 33;

and water company, 61; wheat speculation scandal and, 180–86, 191–92; and wealth and estimates concerning, 78; and would-be assassin, 198

Flood, James L., 145, 149, 185, 191

Florence, Billy, 124

Foster, Ned ("Lame"), 60

Fraser River, 11

Free and Accepted Masons, 14, 98

French Atlantic Cable Company, 164, 176, 178

Fritz, Henry, 131

Fulton, Chandos, 205

Fulton, R. L., 219n41

Garrett, John Work, 164, 166

Garrett, Robert: assistance to Bankers and Merchants' Telegraph, 170; expansion of Baltimore and Ohio Telegraph by, 164; death of father and, 166; description of, 166; erratic decisions, 168, 172; Jay Gould and, 166, 168, 186–87; investments of, 166, 186; Mackay and, 164–65, 168, 172–73; and mental strain, 186–87; J. P. Morgan and, 186; Edward Stokes and, 186, 187; William H. Vanderbilt's death and, 171–72; withdrawal from Mackay alliance by, 166

Geiger Grade, 7, 141

Gem Saloon, 13

George, Henry, 112, 126

Gilded Age, x, 2, 83, 154

Gillis, Will, 56

Gold Bluff, 11

Gold Hill: Bank of California of, 93; as boomtown, 58; Eilley Orum Bowers and, 107; Cornish miners and, 56; development of, 22; diggings of, 27–28; and fire of 1875, 95; Mackay and, 28, 50; mines of, 29; Adolph Sutro and, 51; Virginia City and, 22; as workstation, 59

Gold Hill News (newspaper): on Bank of California suspension, 91; on Con. Virginia mine, 66, 67, 85; on Crown Point bonanza, 66, 225n20; Alf Doten as editor of, 96; on election of 1881, 140; on fire of 1875, 94, 95; on Hale & Norcross mine, 43–44, 45–46; as leading mining journal in Nevada, 45; Mackay and employee of, 86–87; William Sharon and, 222n31; on Sierra Nevada mine, 136–37, 138; on Adolph Sutro, 51

gold rush, 31, 35

Goodman, Joseph (Joe), 53–54, 76, 192

Goodwin, C. C.: on James Fair, 55; and far western history, 3; on Mackay and demeanor, 126; on Mackay and fire of 1875, 94, 98–99; on Mackay and generosity, 98–99, 231n38; on Mackay and William Sharon, 214n17

Gould, Jay: attacks by, 169, 170–71, 173, 180–81; newspapers controlled by, 153, 156; James Gordon Bennett Jr. and, 150, 153, 154, 176, 189; cable breaks and, 166; communications systems controlled by, 153, 164, 188, 189; and competitors, 154, 156–57; ethics of, 155; as financier, x, 150, 154–55, 157–58; Robert Garrett and, 166, 186–87; health of, 188; James Keene and, 156, 160–61; legend of, 155; on Mackay, 190; Mackay and financial attack against, x, 159, 164, 176, 189–90; Mackay and perception of, x, 157, 168, 186; and Mackay reconciliation, 179; monopolistic behavior by, 154–55, 157, 188, 189; newspapers on, 156; *New York Herald* on, 163, 181, 188–89; *New York Times* on, 154, 156, 157, 177; and *New York Tribune,* 156, 180, 183, 189; notoriety of, 154; physical accosting of, 156; Joseph Pulitzer on, 157; Russell Sage and, 162–63; *San Francisco Chronicle* and, 181, 183; Grant Smith on, 181; and son, 188–89; Edward Stokes and, 166, 168; U.S. Senate action against, 164

Gould & Curry mine: Con. Virginia and, 50, 63, 67, 69, 76–77; and diminished prospects, 28; employment in, 113; James Fair and, 99, 103; and 1875 fire, 96; the Firm and, 68, 113; formation of, 25; as Mackay's residence, 99, 102; mine president on, 76; ore body of, 25, 63; worker injury and, 55

Gorham, Harry, 229n7, 237n16

Grand Hotel (San Francisco), 47, 74, 75, 131

Grant, Ulysses S., 33, 120–21, 127, 141

Gray, Elisha, 161

Great Basin, 19, 59

Green, Norvin, 159, 170

Hale & Norcross mine: Bank Ring and, 6, 7, 34, 41; bidding wars and, 42, 43; Bonanza Firm and, 6, 34, 44, 81, 113; Con. Virginia mine shaft building and, 142; James Fair and, 32, 42–43, 44, 45, 62, 99; James Flood and, 43; *Gold Hill News* on, 43–44, 45–46; joint shaft and, 107; John Kelly and, 45; location of, 41; Mackay and, 42–43, 45–46, 104; mills and, 6, 81; miners confinement

and, 42; ore deposits of, 49, 62, 65, 104; perception of, 47, 65; *Territorial Enterprise* on, 43; William Sharon and, 6, 32, 34, 41; stockholders and, 46; yield of, 41, 46, 68
Hallidie, Andrew Smith, 73
hard-rock mining, 2, 24, 26
Harland, Edward, 168, 169, 171
Harpending, Asbury, 87
Hart, Fred, 140–41
Hay-market Riot, 198–99
Hayward, Alvinza, 32, 33
Hearst, George, 25, 76
Hellman, Isaias, 185
Henning, John, 24
Hereford, John B., 81, 91, 102
Heydenfeldt, Solomon, 65, 130
Hirschman, Adolph "Dolph," 27, 36, 71, 74
Hlévy, Ludovic, 121
Hobart, William S., 61, 106, 115
Hoffman House Hotel and Restaurant, 125, 126, 167, 171, 186
Hooper, Lucy H., 121
Howland, Mrs. Robert, 36–37
Hungerford, Ada, 35, 38, 47
Hungerford, Maj. Daniel E., 15, 35–36, 122
Hungerford, Mrs. Daniel, 35–37, 38–39, 74, 122

Imperial mine, fire at, 93
International Telephone and Telegraph, 211

James, Ronald M., 103, 108, 217n5, 218n21, 232n17
Jones, J. P.: Alvinza Hayward and, 33, and Mackay, 145, 192; as a multi-millionaire, 78; Mark Twain and, 192; as a U.S. senator, 76
Juanita, stabbing by, 13–14
Justice Mine Company (Corp.), 135, 140, 226n35

Keefe, James, 25
Keene, James: C. W. Bonynge and, 193; on Comstock mines, 107, 112, 113; feud between James Flood and, 106–8; James Flood on, 107; on James Flood, 33; fortunes of, 86, 113; Jay Gould and, 156, 161; as president of Postal Telegraph, 161; racehorses and breeding by, 161; and rumors, 91; William Sharon and, 84, 86; stock market bear attacks and, 106–8, 112, 160; and Western Union, 160; wheat market and, 160–61
Kennedy, Alec, 17
Kentuck mine: description of, 29; James

Flood, William O'Brien, and, 30, 43; J. P. Jones as superintendent of, 33; Mackay and, 28–30, 38, 52; William Sharon and, 32, 39, 53; Sutro Tunnel and, 52; John Walker and, 28–30, 32, 53, 76
Kern River, strike of, 11
Kinney mine, 65, 75, 130
Klamath County, gold at, 11
Know-Nothing Party, 10

Lake Tahoe, 61–62, 94
Lannen, P. H., 207
Leggo, W. A., 161
Leslie, Mrs. Frank, 23, 54, 104–5
Lewis, Oscar, 3, 106, 122, 127, 220n54, 245n30
Lick House (hotel), 4, 128, 147, 197
Lingenfelter, Richard, 56
Little Big Horn River, 155
Little Gold Hill, 24; mines, 28
Locklin, Wilson, 57
Logan, "Big," 13
Longfellow, Charles, 152
Lord, Eliot: on Comstock mills, 129; description of Comstock by, 224n3; on Hale & Norcross stock prices, 221n20; on James Fair, 45, 144; on Mackay, 27, 126, 144, 221n21; on William M. Stewart, 228n19
Low, C. L., 42
Lyman, D. B., 145
Lyman, George, 29, 219n41
Lynch, Philip, 45, 222n31

Mackay, Clarence, "Claire": birth of, 74; and brother's death, 203; James Corbett and, 196; culture and, 109; and father's illness, 199–200; as businessman, 205, 208, 211; gifts to University of Nevada by, 211; marriage of, 111, 205
Mackay, Eva (Bryant): and broken hip, 36; culture and, 109; death of, 201; Mackay and, 47, 70; marital problems and, 199, 200–201; in Paris, 47, 68; in Virginia City, 36, 37; wedding of, 149
Mackay, John William: actors and, 124; administrative position of, 33–34; Alexander III coronation and, 143; appearance of, 4; assassination attempt against, 197–99; and Aurora, 24; Bank of California's failure and, 91; Bank Ring and, 79; benevolence of, 1, 126–27, 141, 196–97, 207–8; James Gordon Bennett Jr. and, 143, 151, 153, 154, 157, 162, 166, 174–79, 205; James Gordon Bennett

Sr. and 9, 48–49; and Big Bonanza, 78, 88, 97; as "The Big Bonanza," 149; boys and, 103; birth of, 8; Bonanza Firm partnership and, 43, 129; Bonanza Firm and percentage owned by, 3, 33; C. W. Bonynge and, 106, 193–94; George S. Brander and, 182, 184; Louise Bryant and, 36–38; building infrastructure of mines by, 76; and Bullion mine, 27–28, 33, 50; Burleigh drills and use by, 77; business acumen of, 5; business attitude of, 153; businesses of, 205–6; and Caledonia tunnel, 26; and Canadian Pacific Railroad, 178; character of, 45, 54, 126, 210; childhood of, 3–4, 8–9, 48; children and, 122–23; children of, 47; clean money and, 2, 230n3; Dr. A. M. Cole and, 103–4; comments to Dan DeQuille, 86, 87; and Comstock Lode, 15, 27, 53, 67–69, 144–45; on Comstock mines, 141, 144–45; Comstock vein and study by, 65; and concern for others, 97, 98–99; and confrontations with William Sharon, 6–7, 214n17; Con. Virginia mine and abandonment by, 146; on Con. Virginia mine riches, 70, 78; Con. Virginia mine and development by, 67, 71, 144; Con. Virginia mine and investments by, 68, 71; Con. Virginia mine and management by, 69, 76–77, 85; Con. Virginia mine and mining richer veins by, 107; Con. Virginia mine and ownership by, 4; Con. Virginia mine and purchase by, 64–65; on Con. Virginia mine stocks, 71, 87, 89, 104; James Corbett and, 195–96; Crown Point mine and, 64; Sam Curtis and, 65–69; cutthroat tactics and rejection by, 47; Sam Davis on, 1, 4, 81; death of, 210; and death of John Jr., 202–5; Dan DeQuille and, 4, 54, 69, 72, 86; S. P. Dewey and, 114–17; R. V. Dey and assistance to, 199, 200, 210; R. V. Dey and feelings of, 200; R. V. Dey on, 211; R. V. Dey on wealth of, 206, 208; as dominant figure, 78–79; Alf Doten on Fred Hart and, 140–41; Downieville and, 11, 14–15; and employees, 55–57, 62, 88, 206; education of, 4, 9, 27; and Esmeralda claim, 24; on James Fair, 137, 185; James Fair and achievements by, 57, 58; James Fair and assistance by, 140; James Fair and assistance to, 33, 42, 79, 183, 184–85; James Fair and comparisons to, 102, 149–50; James Fair and competition and, 53, 54; James Fair and feelings about, 34, 135, 150, 180–81; James Fair and introduction to, 31; James Fair and relationship with, 57, 134–35, 146, 148–49; James Fair and verbal attacks on, 111–12, 148, 181; Theresa Fair and, 148, 183; family and, 5; father's death and, 9; fighting and, 139, 194; financial assessment and, 139–40; financial warfare and, 150–51, 164; and fire of 1875, 94–99; on James Flood, 192; James Flood and friendship of, 34, 127, 148, 192; James Flood and William O'Brien and introduction to, 30; James Flood and William O'Brien with, ix, 33; James Flood on, 34; Free and Accepted Masons and, 14; friendship and, 53, 126, 134; gambling spirit of, 62; *Gold Hill News* and, 42–43, 45; Jay Gould and, x, 154, 164, 189; Jay Gould and Atlantic cable cartel challenged by, 150, 157–58, 174–79, 186; Jay Gould and attacks against, 169–70, 180–81; Jay Gould and Western Union challenged by, 157–58, 159, 164–66; Jay Gould perceived by, x, 157; Ulysses S. Grant and, 120–21; Hale & Norcross mine and, 42–44, 45–47; habits of, 27; heritage of, 9; heroic stature of, 71; and hired help, 38–39; home built by, 38; as "the honest miner," 3; and Idaho mine, 49; illness of, 199, 200, 205, 210; in-laws and, 38, 121, 122; and Kentuck mine, 28–30; legends concerning, 1, 14, 16, 140–41; Oscar Lewis on, 3; litigation and, 128–33, 244n13; and living in a dugout, 17–18; loss of son by, 6; Eva Mackay and, 200–201; Louise Mackay and, 68, 70, 192, 193–94; management skills of, 45; Fr. Patrick Manogue and, 97–98; marriage of, 5; Metropolitan Opera Company and, 208; and Mexican mine, 20, 21, 50; and mills, 81; and Milton mine, 25, 26; on mining, 66–67; mining development by, 25–26, 76, 144–45; monopolies and confrontations by, 2–3, 150, 157–58, 243n37; and Nevada Bank of San Francisco crisis, 179–86; as Nevada Bank of San Francisco director, 82, 185; newspaper attacks on, 116–18; newspaper support for, 117–18; nobility and opinion of, 5, 8; obituaries of, 210–11; Jack O'Brien and, 14–15, 16, 17–18, 24, 126; William O'Brien and, 128, 229–30n13; and Oregon tunnel, 26; Paris and, 142; Paris Exposition and, 123; as patron of art, 123, 143; and poker, 27, 151; and politics, 150, 154; politicians and, 120; portrait of, 127; Postal Telegraph and, 159–61, 162–173, 178–79, 187–90; and power, 5, 214n13; praise of, 3; prejudice and,

4, 10, 215n7; William Ralston and, 75, 87, 92; residence of, 4; respect earned by, 27; Russell Sage on, 162–63; St. Mary's Church and, 94, 97–98; San Francisco and, 73–75; and scandal fraud, 130–33; and servants, 202, 245n30; William Sharon and, 39, 76, 92; William Sharon and assistance to, 49; William Sharon and competition and, 49; William Sharon and cooperation with, 50, 53, 76; shipbuilding apprenticeship of, 10; Sierra County and mining by, 4, 11, 14–15, 20, 24; Sierra Nevada mine and, 137; Johnnie Skae and, 136; Grant Smith and, 145; Grant Smith's material on, 3; social events and, 49, 121–22, 149; speech impediments, 4; statue of, 211; stock market attacks and, 106–7, 114–16; stock market speculation, 229n7; Edward Stokes and 124–25, 167–69, 172–73, 194–95; and superintending of mines, 31; Adolph Sutro and, 50, 52–53; taxation and, 80–81, 87, 154; telegraph rates and, 3, 189–90; transatlantic cables and, 160, 164–66, 174–78, 189–90, 205; transpacific cable and, 208–10; temper displayed by, 116, 182, 194; *Territorial Enterprise* and, 43, 76, 140; the theater and, 124, 125–26; Tiffany silver and, 123; timber framing by, 4, 20, 25, 26; and travel to goldfields, 10–11, 215n10; tribulations of, 6; Mark Twain and, 192, 201–2; and Union mine, 17, 20, 26; U.S. and gratitude of, 151; U.S. presidents and, 120, 143; and vacation, 68; and Virginia City, ix, 16, 24; J. M. Walker and, 28–30, 33; George G. Ward and, 164; and water company, 61; on wealth, 151; wealth and estimates concerning, 54, 78, 143, 205, 206; wealth and status of, 4; wheat deal and 179–86; work ethic of, 4

Mackay, John William "Willie," Jr.: birth of, 47; as businessman, 200, 202–3; closeness to father of, 202, 204, 208; culture and, 109; death of, 203; and Eva and assistance of, 201; and father's illness, 199–200; funeral for, 204; life of, 202–3; in Paris, 47, 70, 122–23

Mackay, Louise (Bryant) (Hungerford): Ellin Berlin on, 120; Edmond Bryant and, 15, 35–36; Catholicism and, 98, 110; charity of, 6, 74, 108, 120; Clarence and, 74, 109; Don Ferdinand Colonna and hatred of, 201; coronation of Alexander III and, 143; description of, 36, 37, 108; R. V. Dey and assistance to, 109; R. V. Dey and ill feelings of, 199–200; early life of, 34–35; education of, 15; employment of, 36, 39, 110–11; European society and, 119; and Eva, 36, 47, 109; Theresa Fair and, 37, 108, 148; and family, 35, 47, 111, 122; friendship and, 120; gifts to University of Nevada by, 211; grace of, 6; Ulysses S. Grant and, 121; and hired help, 38–39; hospital named for, 108; Queen Isabella and, 142; jewelry of, 111, 142; John Jr. and, 47, 109, 204; lifestyle of, 5, 150; London and, 149, 192; John Mackay and courtship of, 37; and John Mackay's, 210; and John Mackay's, 199–200, 210; John Mackay's marriage to, 5, 37–8, 39, 202; John Mackay's spending on, 207; mansions of, 119–20, 123, 142, 192; mine information and, 68, 70; mother and, 38–39, 120; Nevadans' perception of, 6, 108; news reports of, 119; New York and, 105, 108–11; New York society and connection of, 205; in novels, 121; Paris and, 68, 70, 111, 119–22, 142; portrait of, 143; San Francisco and, 73–75 108–9; slander against, 193–94; social events and, 49, 119–21, 142–43; social status and endeavors by, 5, 37, 109–11, 119, 120–21; spending habits of, 74; suitors and, 36–37; table service of, 123, 142n22; Tiffany silver and, 123, 237n22; Queen Victoria and presentation of, 192; Virginia City house of, 38; withdrawal from public life, 211

Maldonado brothers, 20

Mallon brothers, 95

Mancinelli, Luigi, 208

Mansfield, Josephine "Josie," 124–25, 235n17

Mansfield v. *Fisk,* 125

Marlette Lake, 62

Manogue, Fr. Patrick (Bishop Manogue): description of, 97; and fire of 1875, 98; and Theresa Fair and divorce, 147; and John Mackay's generosity, 98, 231n38; Louise Mackay and, 36; John and Louise Mackay and marriage by, 37; and *Territorial Enterprise,* 141

Marquess of Queensbury Rules, 195

Marysville, 12

Mason, Col. N. H. A., 103

Masten, N. K., 89

Mathews, Mary McNair, 38–39

Mayre, George, Jr., 33, 44, 78, 137

McAllister, Hall, 182–83

McAllister, Ward, 109–10

McCullough, John, 124, 148
McEwen, Arthur, 26, 31
McLane, Louis, 82, 144, 149, 228n29
McLaughlin, Pat, 15
Meissonier, E., 143, 193
Metcalf, Dr., 187
Metropolitan Livery stables, 93
Metropolitan Opera Company, 208
Mexican-American War, 35, 82
Mexican mine, 20–21, 24, 50, 55, 113, 144
Michelson, Miriam, 222n34
Mills, D. O., 27, 89–90
Milton mine, 25, 26
Miner's Union, 138
Modjeska, Helena, 55
Montague, Henry, 124
Montgomery Street, 4, 33
Mooney, Mrs. William, 37, 220n7
Moore, J. S., 177–78
Morgan, J. P., 109, 186
Morgan Mill, 24, 97
Morrison, George Howard, 148
Morrow, Robert "Bob," 25
Mount Davidson, 16
Mount Rushmore, 211
Mount Whitney, 16

Native (Indian), 2, 12, 19–20, 155
Nevada Bank of San Francisco: Bank of
    California loan and, 183; buyout of, 148–49;
    James Corbett and, 195; James Fair and,
    148–49, 183; incorporation of, 82; fight in,
    194; James Flood and, 144, 148–49; Mackay
    and, 139, 148–49, 207; Mackay and buying
    out partners, 168; Mackay and rescue of,
    145, 179–85; Louis McLane and, 82, 144,
    149, 228n29; merger between Wells Fargo
    Bank and, 185; and notes kiting, 179; open-
    ing of, 87, 93; and painting of Mackay, 127;
    sale of, 185; and wheat scheme, 181–85
Nevada Block, 87, 137, 206
Nevada State Journal, 82, 88, 107
Nevada Supreme Court, 25, 40
Neville, Amelia Ransome, 4, 74
New York Herald: James Gordon Bennett Jr.
    and, 9, 48, 143, 151, 152, 177, 189; Jay Gould
    and, 150, 153, 156, 163, 172, 181, 188–89; on
    Mackay, 5, 208, 209, 211; Mackay support
    by, 177, 181; power of, 153
New York Times: on Commercial Cable, 177;
    on Judge Donahue, 171; on James Fair, 147,
    185; on James Flood, 184; on Jay Gould,

154, 156, 157, 175, 177, 185, 187; on Louise
    Mackay, 193; on Mackay's assassination at-
    tempt, 198, 200; on Mackay and discrimina-
    tion, 116; and Mackay eulogized, 211; on
    Mackay's financial loss, 179; on Mackay as
    a young man, 7; on Postal Telegraph, 161;
    on Johnny Skae, 138; on telegraph war, 165,
    169, 170, 171, 179, 187–88; on wheat scheme,
    182, 183, 184
New York Tribune: on Bonanza Kings, 88; and
    cable rates, 153, 189; Jay Gould and, 156,
    188–89; Mackay and accusations by, 181,
    183; on Mackay fight, 194; on Mackay silver
    service, 123; Whitelaw Reid as editor of,
    156; San Francisco Chronicle and, 183; on
    Virginia City, 94; on wheat scheme, 180–81
Ninth Legislature, corruption in, 41
Nob Hill, opulence on, 108
North End mines, 136, 144
Northern Paiutes, rescue of girls by, 19
Noyes, Mrs. Edward Follansbee, 120
Nye, James, 49

Obiston, Frank F., 96
O'Brien, Jack, 14–15, 16, 17–18, 24, 126
O'Brien, William Shonessy: and Auction
    Lunch Saloon, 30, 52, 54; Bank of California
    failure and, 90, 92–93, 230n13; Bonanza
    Firm, 3, 33, 43, 127, 129–30; brokerage
    office of, 33; J. Ross Browne on, 105; and
    Comstock Lode, 64–65; Comstock vein and
    study by, 65; and Con. Virginia mine, 65,
    130; death of, 34, 128, 149; Squire P. Dewey
    and, 114–15, 130; James Fair and assistance
    by, 33; and James Fair on, 54, 105; family of,
    127, 128, 149; James Flood and, 33, 34, 127,
    128; and friends, 128; and Hale & Norcross
    mine, 44; health of, 128; and Kentuck mine,
    43; and litigation, 114–15, 130; Mackay and,
    52; and Nevada Bank of San Francisco, 82,
    149; personality of, 30, 34, 127; success of,
    127; and water company, 61; on wealth of, 78
Occidental mine, 37, 220n7
O'Grady, Alexander, 5, 8, 38, 122–23, 196, 210
O'Grady, Alice, 5, 122, 196, 200, 210
Old Red Ledge, 24
Ophir mine: Lucky Baldwin and, 85; Bonanza
    Firm and, 139; Comstock and original ore
    strike in, 17; and Con. Virginia mine, 63,
    67, 84; Phillipp Deidesheimer and, 21; and
    depression, 28; James Fair and, 32; and
    fire of 1875, 94–98, 231n39; James Keene

and, 86; Mackay and, 97–98, 139, 144; and Mackay's original dwelling, 18; and market excitement, 136; and Mexican mine, 20; D. O. Mills and, 89; William Ralston and, 86, 89; and shafts, 24, 144–45; William Sharon and, 84–85, 86, 89; and Sierra Nevada mine, 135–36; and soil, 21; stocks and, 85–86, 89, 113, 118; and Union mine, 17; and water, 20, 67; yield of, 24

O'Riley, Peter, 15

Ormsby, William, 19

Pacific Mill and Mining Company, 81, 139

Pacific Postal Telegraph Company, 178, 182

Palace Hotel, 75, 128, 131; Mackay and, 197–98; William Ralston and, 75, 89, 93; William Sharon and, 93; Johnny Skae and, 138

Paris Exposition, 123

Parrot's Bank, 43, 44

Patton, W. H., 145

Paul, Almarin B., 17

People's Tribune, The (newspaper), 41

Perkins, George C., 209

Petaluma Mill, 28, 50

Phelan, James, 29

Phil. Sheridan Mining Company, 96

Piper's Opera House, 103, 139

Platt, Edward C., 205

Postal Telegraph Company: and bonds, 166; and Canadian Pacific Railway, 178; and Canadian Pacific Telegraph, 178; Commercial Cable and, 205; franking privileges and, 163, 187; Jay Gould and, 181; government regulation and, 188; Elisha Gray's system and, 161; James Keene and, 161; Mackay as director, 162, 164, 166–67, 170; Mackay and proposed merger of, 172; Mackay and purchase of, 159–60; Mackay and refinancing of, 167; John Mackay Jr. and, 203; naming of, 160; rates of, 162, 165; and rebates, 163, 187–88; George Roberts and, 160, 165; service of, 188; Edward Stokes and, 167; Western Union against, 162, 164, 172, 187; Western Union and rates of, 187–88

Potasi mine, 25. *See also* Chollar-Potasi

Pulitzer, Joseph, 156, 157

Ralston, William: and Bank of California, 75, 87, 88–91; Bonanza Firm and, 87, 92; and California development, 75; and Comstock, 39, 91; death of, 79, 91, 105; and financial

failure, 90–92, 93; James Flood and, 90, 92–93; as hero, 75, 92; Mackay and, 47, 75, 91–92; Mackay and Flood as friends of, 87; as millionaire, 78; newspapers and, 91–93; William O'Brien and, 93; and Ophir mine, 86, 87, 89; and Palace Hotel, 75, 89, 93; and San Francisco society, 74–75; William Sharon and, 39, 43, 75, 91–93; and Union Mill and Mining Company, 39; and the West and confidence, 75, 105

Reid, Whitelaw, 156

Resnick, Judge R. S., 148

Richthofen, Ferdinand Bacon, 64, 224n16

Rippey, W. C., 198

Roberts, George D., 112, 157, 160–63, 165, 168

Rockefeller, John D., 84, 109

Rooney, Theresa, 32. *See also* Fair, Theresa

Roosevelt, Theodore, 199, 209

Rosener, Harry: Louise Mackay and, 36–37; Mackay and 36–37, 157; as Pacific Postal Telegraph manager, 178, 182; as Postal Telegraph manager, 167, 187; telegraph cables and, 157, 182

Rosener, Sam, 199

Rosenfeld, John, 181–85

Rossiter, Raymond, 89

Rue Tilsit, (Paris street), 111, 119

Sacramento Record (newspaper), 160

Sacramento Union, 160

Sage, Russell, 162–63

San Francisco Alta: on Bonanza Firm, 117–18; on Mackay, 117; William Ralston and, 92–93; Adolph Sutro and, 21; on telegraph service, 160; and Western Union Telegraph, 160

San Francisco Bar, 5, 122

San Francisco Bulletin, 91–92, 104, 130, 132–33, 136

San Francisco Chronicle: on Bonanza Firm, 92, 112, 193; on Centennial Exposition and Bonanza mines, 106; S. P. Dewey and, 112, 116; on James Fair, 53; on Fair versus Flood and Mackay, 150, 184–85; James Flood and, 92, 97, 184; on Con. Virginia mine, 77; on fire of 1875, 95–97; Mackay and, 97, 116, 181, 182–83; Mackay scandal and, 131; on millionaires in the West, 78; on mine stock-jobbing, 89; newspapers against, 118; on William Ralston, 92–93; as reformist paper, 112; on Sierra Nevada mine, 136; and Western Union Telegraph, 160

San Francisco Evening Post, 116, 160

San Francisco Examiner, 1, 147, 192, 198, 211
San Francisco Herald, 160
San Francisco Mail, 116
San Francisco News Letter, 116–17
San Francisco Post, 116
San Francisco Stock Exchange, 28, 84, 92, 106
Santa Rita (tunnel water stream), 60
Savage mine, 25, 41, 96, 107, 113
Schussler, Hermann, 61–62, 71
Scott, Winfield, 15, 35
Selover, Maj. J. R., 156
Sharon, William: Lucky Baldwin and, 85; and
    Bank of California, 6, 39, 90–93; and Bank
    ring, 27, 39–41, 44, 49, 81; and Belcher mine,
    65–66, 80; and Bonanza Firm, 34, 43, 65,
    68, 80–81, 97; Comstock dominated by,
    79; Con. Virginia mine and, 65; courts and
    control by, 40; description of, 7; employees
    of, 113; James Fair and, 32, 44, 50; James
    Flood and, 87, 97; Gold Hill News and,
    45–46, 140; and Gould & Curry mine, 68;
    government and control by, 41; and Hale
    & Norcross mine, 42–44; James Keene
    and, 84–86; and Kentuck mine, 32, 39, 53;
    machinations of, 39–40, 85–86, 89; Mackay
    and, 47, 49–50, 86, 53, 76, 92; Mackay chal-
    lenges to, 81–82, 87; Mackay confronted by,
    6–7; as millionaire, 78; and mills, 39, 52, 97,
    106; and monopoly, 40; and Ophir mine,
    84–86, 89; opposition to, 41; perception
    of, 80; and poker, 27; politics and control
    by, 80; power of, 39; William Ralston and,
    39, 75, 86, 91–93, 230n19; and social affair,
    49; stock manipulation by, 58, 90–91, 117;
    against Adolph Sutro, 50, 52–53; and taxes,
    40–41, 80, 87, 128; Territorial Enterprise
    and, 46, 76, 117, 128, 140; and Union Mill
    and Mining Company, 39–40, 59, 117; and
    U.S. Senate, 76, 80, 140; and Virginia &
    Truckee Railroad, 52, 81; and water, 59; and
    water company, 61; and Yellow Jacket mine,
    40, 46–47, 130
Shaw, John B., 55
Shay, Kate "Crazy Kate," 94
Sheridan, Col. Mike, 94
Sheridan, Gen. Phil, 94
Sherwood, Robert, 137
Sides, R. D. "Dick," 225n18
Siemens and Brothers, 160
Sierra County: commerce in, 12; creation of, 11;
    49ers and rush to, 12; Mackay and, 4, 11–12,
    24; and Mackay leaving, 15, 16, 18; mining

wealth of, 14; population of, 12; stories of
    Comstock Lode and, 15; traffic in, 12
Sierra Nevada: and alarm spreading, 19;
    Bonanza Firm flume and, 101; both sides
    of, 106; crossing of, 18, 22; forests of, 22;
    Mackay and, 12; peaks of, 16; and storms,
    28; Washoe east of, 15; water and, 61, 70–71,
    73
Sierra Nevada mine: Bonanza Firm and, 139;
    and broken hearts, 138–39; James Fair and,
    137; and fight, 139; James Flood and, 137;
    location of, 136; Mackay and, 137, 144; ore
    of, 136–37; Johnny Skae and, 61, 138; stock
    of, 136–38
Sisters of St. Dominic, 35
Skae, Johhny: death of, 138; and party, 136;
    personality of, 136; and poverty, 138;
    serenade of, 137–38; and Sierra Nevada
    mine, 136, 138; water company and, 61
Smallman, Amelia Hodgden Fritz, 131–33,
    236n36
Smallman, William H. M., 131–33
Smith, Fred, 53–54
Smith, Grant: on Bonanza Firm, 129; on
    Comstock and rush, 218n21; on Dan
    DeQuille, 54; on S. P. Dewey, 233n31; on
    James Fair, 31, 34, 54, 106, 242–43n29;
    on Jay Gould, 181; on George Lyman
    and Mackay, 219n41; and Mackay, x, 3,
    27, 218n25; on Mackay and Catholicism,
    231n38; on Mackay and Fair, 129–30, 220n7;
    on Mackay and Flood regarding Ralston
    and Sharon, 87; on Mackay's character, 45;
    on Mackay and Stokes, 125; and meeting
    Mackay, 3, 145, 237–38n34; on people of
    Nevada and Mackay, 6; on San Francisco
    Chronicle managers, 233n31; on Sierra
    Nevada mine, 138
Smith, W. H., 139
Société du Cable Trans-Atlantique Français,
    153
Soit, Kinney, 17
Southern Pacific Railroad Company, 146, 206
Sprague Elevator and Electrical Works, 206
Spreckels Sugar Company, 206
Spring Valley Water Company, 62, 89
Stanford, Leland, 78, 108
Stanley, Henry M., 152
Stephens, Mrs. Paran, 110
Stewart, Joe, 134
Stewart, William M., 24, 53, 79–80
St. Louis Globe-Democrat, 148, 149–50

St. Mary Louise Hospital (Virginia City), 108
St. Mary's Catholic Church, 94–95, 97–98
Stock Exchange, The (newspaper), 111, 117
Stock Report, (newspaper), 117, 118
Stokes, Edward, S.: description of, 124, 167; Jay Gould and, 169, 172–73; and Hoffman House, 167, 186; and killing of Jim Fisk, 124–25; Mackay as benefactor to, 167; as Mackay confidant, 167; Mackay and correspondence to, 168–69, 172, 173; as Mackay's friend, 124, 125, 134; as Mackay's partner, 166–69, 171, 172, 187–88, 194–95; Josie Mansfield and, 124–25; murder trials of, 125; prison and, 125; suit against Mackay by, 194–95; Western Union against, 169, 171–73, 187
Storey County, 40, 80, 228n25
Sullivan, John L., 195
Sullivan Mill, 46
Sutro, Adolph: and Comstock Lode vein, 64; Alf Doten on, 51; and Eastern capital, 51; Mackay and, 49, 50, 52–53; and mining system, 21, 50; mistakes of, 51; William Sharon and, 52; and tunnel, 50
Sutro tunnel: Bank Ring against, 52; and government charters, 51; mine subscriptions pledged to, 50; William Sharon and, 40, 50–52; superintendent of, 32; Adolph Sutro and, 21; and Virginia & Truckee Railroad, 53

Tammany hall, 125, 168
Telfener, Count and Countess, 142
Territorial Enterprise: on extent of Big Bonanza, 85; Bonanza Firm defended in, 118; on Con. Virginia, 69, 72, 85; Con. Virginia mine bids in, 65; on Con. Virginia mine formation, 63; Dan DeQuille in, 63, 69, 72; Alf Doten and assistance to, 96; influence of, 23; on James Fair, 43, 135; on fire of 1875, 94–95; Joseph Goodman and, 53–54, 76; on Hale & Norcross mine, 43; Fred Hart and, 140; on Lake Tahoe water, 61; Mackay and, 18, 43, 94–95, 117, 140–41, 142; on Louise Mackay, 120; on Louise's marriage to Mackay, 37–38; on mining and milling, 117; Bishop Manogue and, 141; William Sharon and, 46, 76, 117, 128; on slander against Mackay, 132; on Adolph Sutro and new town, 51; on tax laws, 128; on Yellow Jacket mine, 46
Tevis, Lloyd, 25, 76

Thomas, C. C., 32, 42
Townsend, Dwight, 170
Tribunal of Commerce, 178
Turner, Frederick J., 2
Twain, Mark, 32, 181, 192, 201–2
Tweed, William M. (Boss), 83, 168, 171

Union Consolidated mine, 17, 20, 26, 136–38
Union Mill and Mining Company, 39–40, 59
Union Pacific Railroad, 83–84, 156, 164
United Lines Telegraph Company, 169, 171–72, 187–88
United Telegraph Lines, 165, 169
Utah mine, 113, 198

Van Bokkelen, Jacob L., 108, 232n17
Vanderbilt, Consuelo, 202
Vanderbilt, Cornelius, 83, 109
Vanderbilt, William H., 152, 171
Vanderbilt, William K., 111
Virginia and Gold Hill Water Company: Bank Ring and, 60; board members of, 61; capital stock of, 60; Dan DeQuille on, 70; the Firm and, 129; formation of, 60; and litigation, 60–61; Mackay and, 61, 129, 139; Hermann Schussler and, 61–62; William Sharon and 40, 61; Johnny Skae and, 136; and water pipeline, 61–62, 70
Virginia and Truckee Railroad: Bank Ring monopoly and, 53; construction papers filed for, 52; and Con. Virginia mine, 99; and fire of 1875, 96, 99; fraud by, 53; William Sharon and, 40; Adolph Sutro and money appropriated by, 40; taxes and, 80–81
Virginia City: Aurora and, 24; Bank of California and, 91, 93; Bank Ring and, 2; J. Ross Browne on, 19, 22, 59; as boom-town, 58–59; C. W. Boynge and 193; C. W. Boynge on Louise Mackay and, 193; Bullion mine and, 27; Comstock Lode mine, 2; Con. Virginia mine excitement and, 77–78; Philipp Deidesheimer and, 21; and depression, 28, 88; description of, 22–23; ethnic communities and, 56; James Fair and, 32, 42, 102, 135; fame of, 78; and fire of 1875, 94–99, 113; Hale & Norcross mine and, 46; Dolph Hirschman and, 74; hostile environ-ment of, 74, 227n2; Louise Mackay and, 5, 15, 36, 74, 108; Mackay and, ix, 1, 4, 16, 24, 38, 98, 102–3, 118, 133, 137, 145; Mackay and mines of, 140, 144, 145; Mackay and theater

of, 124; Mackay's friends and, 182; Mackay shooting and, 197; naming of, 17; and mines, 118; Paiute Lake War and, 19; Judge R. S. Resnick and, 148; George Roberts and, 161; St. Mary Louise Hospital in, 108; William Sharon and, 7, 43; social event in, 49; William Stewart and, 79; stockholder riot in, 138; Adolph Sutro and, 51, 52; and *Territorial Enterprise,* 117; violence in, 74; Old Virginia Finney on, 17, 217n5; and water, 60, 61, 70, 73

Virginia Evening Chronicle, 26, 40, 96, 139

Walker, J. M., 28–30, 32–33, 53, 76
Walsh, James E., 205, 208
Ward, George G.: and cable landing, 175; on cable rates, 176, 179, 186, 189; and Canadian Pacific Telegraph merger, 178; description of, 164; Mackay and, 189, 210; Mackay hiring of, 164;
Washington saloon, 95
Washoe County, 15, 140
Washoe Indians, 15, 70
Washoe Valley, 61–62, 101
Watson, "Uncle Billy," 44
Webb, William H., 9, 10
Wells Fargo & Company, 25, 76
Wells Fargo Bank, 91, 185
Western Union (Telegraph): and American Rapid Telegraph, 168, 169–70; and Associated Press, 160; and Baltimore and Ohio Telegraph Company, 186; bear market and, 156; cable rates and, 176; Albert Brown Chandler and, 187–88; and competition, 164–65, 172, 186; Congress and, 163; earnings decline, 172; far West

newspapers and, 160; formation of, 159, 170; as first nationwide multiunit business, 159; franking privileges by, 163, 188; Norvin Green and, 159, 170; Jay Gould and control of, x, 153, 156–57, 173, 188; Elisha Gray and, 161; James Keene and, 156, 160–61; losses of, 189; Mackay and confrontation of, 2, 165, 167–68, 176; monopolization by, 2, 157; Pierpont Morgan and, 157, 186; Postal Telegraph competing against, 162, 165, 172, 187–88; and railroad line rights, 163; Russell Sage and, 162–63; stocks of, 179; Edward Stokes and attacks by, 171; Edward Stokes and suit against, 172–73; strike against, 162; and telegraph rates, 165, 172, 188–90; train depots and offices of, 163; triumph of, 171, 173, 188, 190; U.S. market controlled by, 156; U.S. Senate and, 156, 163–64
White, James, 114, 116, 118
White and Murphy mines, 63, 65
Whitehill, Henry R., 70–71
whites, 2, 12, 19–20, 36
Wood, Billie, 49

Yellow Jacket mine: description of, 29; and 1871 development, 61; employment in, 30, 113; the Firm and, 113; James Flood and, 144; location of, 29; Mackay and, 47, 144; superintendent of, 41; William Sharon and, 39–40, 46, 47, 130; stockholders and, 40
Yerington, Henry M., 59, 140
Young, John Russell: on cable war, 189; Mackay and, 3, 151, 179, 196; old western history and, 3; on partnership of Mackay, Flood and Fair, 242n27
Yuba River, 11, 12